# THE QUANTUM
# CONNECTION THEORY

# THE QUANTUM CONNECTION THEORY

Chaz C. Disly

**To order additional copies of this book, contact:**
Xlibris Corporation
1-888-795-4274
www.Xlibris.com
Orders@Xlibris.com
21262

# CONTENTS

to the future generations.

# INTRODUCTION

Have you wondered why some people work so hard yet have so little? Why do some people have so much yet work little if at all? Why are good people being stepped over or, on many times, the bad guy comes out smelling like a rose? Why is it possible for people to do dishonest things, and not be exposed? Do not be too quick with your answer.

One of the most perplexing things to me as a child was seeing my religious leader, married and with two lovely daughters, teaching the Ten Commandments to us while using the congregation as his personal harem.

Albert Einstein once said that we use 10 percent or less of our brain. What, then, is all the extra brain capacity sitting there for? Have you ever had a good idea, which you never told anyone about, and yet, some years later you see that someone else has, what you thought was your idea, on the market?

Have you ever watched an event occur and knew it would happen before it actually did? Did your prediction cause the effect? Well now, indulge me as I offer some possible insight into these and many other puzzling thoughts.

This is not a book to bash religion. On the contrary, I recognize religious values and how it has provided continuity in bringing about our modern human experience. Our great society has risen through the principles and teaching of many religious texts. Do not assume that this book is intent on devaluing religion. The prime objective of this book is to gently nudge humanity back onto the mental plan we all once knew, as well as the longevity we once enjoyed before the Great Flood of our planet.

Because of the many different beliefs we share, let us, when reference is made to that of which we hold as Supreme and Most

High, use this symbol ( ) to insert that name. Most of all, never loose sight of your personal ( ) concept.

With my writing comes a strong warning: a misuse of the kind of knowledge in this book can be disastrous to you and your loved ones. Because quantum physics is based on the functioning of subatomic particle charges, your thought waves, negative or positive, contribute to and make a difference in the order of things within our universe.

For years, I debated on whether or not I should write this book. At first, I started making notes built on notes I received from one of my uncles that I was to pass down to a future thinker of my family tree as well. Just as the chosen family member from his past generation had chosen him to pass what they felt should be kept secrets, down to him, I was chosen by this uncle, a year before he died, to carry these notes forward.

However, I have now come to believe that we are all a part of the same family tree and that humanity is best served when everyone is able to share in these ideas. The information I will put forth is not new, a handful of thinkers throughout our civilization have always been aware of it. I guess some people, like me, just cannot keep a secret.

Now that we, the human race, have started studying the quantum world, I feel the time is right to empower those who are ready with the knowledge of the Quantum Connection Theory. If you have or can elevate yourself above hate, envy, fear, selfishness, anger, and most of all prejudice, then you are ready. Anyone is ready if they are willing to do their own thinking. To get anywhere with this Quantum Connection Theory, one must be willing to have self-reliance in their own reasoning, instead of the clichés and the habitual ways of thinking, which were passed down to them.

We are capable of amazing things, and this book will show you how to make some of those things happen. However, when one is able to open their imagination and do their own thinking, it becomes apparent what Albert Einstein meant when he said, "Imagination is everything."

Paraphrasing Jesus as he once said, "These things you see me do, you will do and more." If you can believe this teaching of Jesus, then simply take him at his own words. Start doing that "more" in which Jesus assured us of. There have been people throughout the past who, when they stumbled on the "more," were severely scolded for demonstrating what could not be understood by the masses.

I see no reason why science and religion cannot coexist. There is no need for the argument about creation versus evolution. To better understand creation, we must first acknowledge and understand that there are physics involved. The ones who believe in creation must acknowledge the physics in the creation. The ones who believe in evolution and the big bang must acknowledge the miracle of life that is inherent in the creation of humanity (Hint: The Quantum Connection).

I have chosen to write in a short-paragraph style to give readers a chance to think about every thought I present. In each little paragraph, though it may not always really support a full thought on its own, the paragraph breaks allow one's brain a chance to process the input. However, most of all, the breaks allow the input to circulate through one's memory and consciousness, stimulating a mental dialogue within. The paragraphs are more like a thought-provoking bit for one to think about as they visualize each scenario.

The stories I will share with you are not to uproot your beliefs or cause external debates; they are only there to bring transparency to preconceived notions. "Imagination is everything"; we should therefore do all that we can to enhance our imagination. The stories I will share with you are to stimulate and dilate your imagination for the outcome of this book, which will be your quantum connection.

Because the quantum connection must be personal within each of us, one must rely on his or her own imagination in creating his or her universe. Not even this book should restrict one's creative thoughts. Restriction should come only from the extent of one's imagination. You will find stories in the book that may stretch your imagination to its limit. Nevertheless, stretching one's imagination is precisely what must be done if one is to truly perfect their quantum connection.

Even though I have kept this book small enough to fit into your pocket, you will find that it is big enough to last a lifetime. Enjoy your thoughts as you take time to think your way through this little book.

# CHAPTER 1

## Humble Beginning

As a little boy growing up in the early forties in the seaport town of Savannah, Georgia, I became curious about the universe and ( ) *see introduction,* at a very early age. I started asking questions about my thoughts as early as four years of age. I wanted to know how thoughts work. However, my daddy did not like answering my questions. I can see now that I was just asking questions that my parents could not answer. Still, I kept asking. My inquiries were usually distressing to my dad, and the few times he did try to give me some answers about ( ) and the universe made me more confused because they just never quite made sense to me. Like most kids my age, I would ask my father to explain about the stars and moon. I remember, he somehow thought we lived inside the earth, and that the earth was the center of the universe.

I remember in 1943, during World War II, we would usually go for a family drive on Sundays. Daddy, not being in the war, would pack us all up on Sunday morning into our family car, which Daddy called "Rags," and we would drive to the country to visit my grandmother. My older brother Henry was then six; my younger sister Inez, three; and myself, four all shared the backseat with Inez in the middle, and I always sat behind my mother on the right. Rags was a dirty black four-door 1936 Ford with tattered-and-torn seat coverings. The roof of Rags had an interior cloth that would sometimes fall, and Daddy would always find a way to glue it back up.

My grandmother, on my mother's side, was whom we visited. She lived south of Savannah about fifty-two miles in the small town of Eulonia, Georgia, which was a two-hour trip for Rags only because Dad, being a cautious man, never went over thirty-five miles per hour. Even so, our tires would get hot, and we were sure to have at least one or more flats on each trip. Daddy never allowed us to eat peanuts in the car, as he was convinced that it jinxed Rags into having a flat tire.

What I loved most about these trips was that I had Dad as my captive, but not necessarily captivated, audience for my questions. I remember one Sunday evening as we were returning to Savannah, I tried getting Daddy's attention, but he said, "Boy, can't you see I'm thinking?"

I replied, "How do you think, Daddy?"

Dad gave me a quick glance, with a disgusted look and a frown on his face, in his rearview mirror. "How do I think? With my mind, stupid," he replied.

Not being discouraged, I replied, "Yeah, but how do thoughts get in our minds and work? You know what I mean, Daddy?" Poor Daddy got so angry with me; he roared at the top of his voice, "Boy, quit asking me all of those *stupid ass questions!*" I fell back into my seat and did not ask anything else for the remainder of that trip.

"Stupid" was Dad's most frequent adjective when referring to me. Still, I would ask questions about ( ) and the universe, which he or no one else in our circle of friends, at the time, could give any sufficient answers to. The stage was being set for a long life of intriguing curiosity, "How do our thoughts and minds work? How does ( ) work in us?"

The people in the circle I grew up around had always thought me to be strange in my thinking. Perhaps if someone had answered my questions to my satisfaction, I could have moved on to whatever normal people spend their time thinking about too.

So here I am now, at the age of sixty-five, still trying to answer those *stupid ass questions* or, should we say, *seriously asked questions* (saq). In this book, I will share some of my conclusions thus far during my lifetime here on this earth.

In some circles, it's thought that we choose the family we find ourselves being born in. It's further believed that we choose our parents to which the purpose of our life experience can best be fulfilled. One thing is for sure: knowingly or unknowingly, we may be the problem as well as the solution to the problems we are trying to solve. Some of us are the problem for others, and some are the solutions to their problems. We are, therefore, all vital to the solution. I have come to the realization that the problems in problems are no problem, once the problem is understood.

Through many years of mediation and some help from the reading of Carl Young, I have learned and concluded that there is a Universal Consciousness (UC). All that we think, have thought, and will think is stored in, and processed through, this universal mind. It holds the essence of our discoveries and experiences. It reacts to the development throughout our human span. This UC is not restricted by time—past, present, or future.

Individuals like me cannot let go of such perplexing SAQ until we get an adequate answer. The questions will not just go away. We just keep asking ourselves for the answer. Therefore, from the universal perspective, we will answer the question adequately, or it will not go away. We can neither ignore nor forget the question. We must solve and answer the questions that haunt us.

My dear old dad had no clue about unanswered universal questions. I guess my dad and people like him have the task as givers of the problem. Dad spent most of his life never knowing that there was a problem or that he may have been a part of, if not the problem itself.

In spite of what I have previously said about my dad, I have to admit that he must have been a bit clever when in 1942 he got drafted into WWII; you may want to ask, how is this clever? Well, my father hated and was afraid of guns; he did not care for police action or airplanes, and he definitely did not want to be out on the water. He was just plain scared of war.

When the army drafted Dad that winter, he wanted out and fast. When it was his day to clean the latrine, he put a bit of octagon soap in his mouth to create the illusion of foaming at the mouth and pretended to have seizures. After that, he started to cry on a whim and pretended that he was too stupid to follow directions from anyone. Three weeks after being drafted, he was sent home as a useless, 4F crazy man, unfit for military duty.

When Daddy knocked on the door early one morning about 5:30 a.m., my mom was so shocked. We all got up and shouted, "Daddy, Daddy!" However, Mom just spent the next two hours or more crying. Five decades later, just before she died, my mother admitted to me that she cried so hard because she thought she was free of him.

I decided that Dad was put here on earth as the developer of my extraordinary low self-esteem. What's more, he did his job superbly. Allow me to explain.

Dad was an uneducated man who supported his family by working as a laborer, a handy man. I must say that he was quite handy. I came into his world as the second of six kids. My older brother was named, like most first-born are, after Dad and was the apple of Dad's eye. I, on the other hand, was named Charles after my mother's old lover. Who, I should add, secretly continued an affair with her even after I was born.

Dad simply never believed that I was his, so he hated me. My guess is that he could not put me out of the family because I was Mother's love and he loved her. As a result, I became the recipient of an enormous amount of verbal and physical abuse from him.

Nevertheless, do not frown upon my dad. Now, I have empathy for his peculiar circumstances. I guess he was just doing his job, developing my low self-esteem. If it is that I came to earth to solve some spiritual problems, then perhaps, I needed that kind of childhood to prepare me for my journey through my lifetime on earth. My dad has since passed away, and the pain has long been gone. I want to thank him for a job well done.

As we bicker and blunder through life solving our problems, the UC develops. Although we have tried many forms of governments in our constant search to find the best system that will accommodate the human experience, still we have obviously fallen short. When we finally get it right, the basic needs of all humans will be met, and there will be no need to rebel against anything.

When we get it right, there will be no terrorist attacks, no more wars, no more distrust. As we continue to solve our problems, a workable government will emerge for the entire planet. A free market under a democratic government does seem to provide the best opportunity for industrial progress. It also seems to provide the most freedom for the human spirit. We must continually seek improvement, until everyone can share in and enjoy this freedom of the human spirit—and I mean every human being on earth— there will obviously still be things we must discover in governmental achievements. We must learn to think beyond our personal place mats and think about the world as a whole, considering the needs of our very fragile planet. We should learn that unselfishness could be quite profitable. All situations should be that of a win-win situation.

I have met people who could not buy into the notion of a UC. If you happen to be one of those folks, allow me to put forth another perspective. Let's assume that you have more knowledge than your parents and grandparents, and so therefore, your children and grandchildren will have even more knowledge than you. This process is being accomplished through an accumulation of knowledge that is passed down through the generations. We record in textbooks, as well as the cognitive knowledge is increasing through the complete human race, then consider this as the universal consciousness.

As individuals, we pass away leaving our understanding and knowledge, of the world we lived in, to the next generation. One could ask, "What was it all for if we ourselves must pass away?" Consider this, the universe may have needed a way to experience consciousness and therefore has raised us up as its consciousness, its identity, its way of understanding and experiencing itself. The universe is seeking comprehension of itself through us, we being the physical expression of its desire. After all, we are but universal matter set forth to ask questions about ourselves as well as anything else we become aware of.

As we understand it, we catalog and transmit our findings and solutions unknowingly, out into the universe. This information then becomes part of the universal consciousness. The universal consciousness is all knowing, and yet it is nourished, among other things, by our consciousness. By calming our senses and entering into a meditative, trance-like state, we are able to tap directly back into this UC.

Some of us have been more proficient in tapping into the UC, such as Edgar Cayce. However, we all have this ability, and with some inner searching and meditating, this ability can present itself in all of us in varying degrees.

Edgar Cayce was born on 18 March 1877 and made his name in the first half of the twentieth century in America as a psychic healer—perhaps the greatest that the United States ever produced. During his lifetime, he was credited with assisting thousands of people suffering from all manner of ailments. (www.edgarcayce.org)

# CHAPTER 2

## Cognitive Dissonance

The word "belief" is by far one of the most honest words I have ever found in the dictionary. Be-lie-f. There, it tells us directly to *be* aware of the *lie*. Belief is the word "belie" with an "f" added to it to remind us that what we hold as truth today may become future lies. Like the world is flat. So with that in mind, let's be willing to have a little fun with our beliefs. A thousand years from now, people will laugh at the foolish lies we once believed.

Cognitive dissonance is the psychological confliction results from incongruous beliefs and attitudes held simultaneously. I, like most children growing up, believed that my parents knew everything. It is not until later that we learn to question what we learned. I waited until I was twenty-eight before I really thought to verify everything for myself.

I mentioned earlier that I had many questions about life and the universe. I was never quite satisfied with the answers from my father and other adults around me. But somehow, I never truly question my teaching. Most of all, the one thing that I did not question was the fact that I was "stupid." I heard it so many times from the one person whom I trusted for truth; I just started to accept it. I began calling myself stupid.

The childhood data I had in storage, "I am stupid," was so deeply imprinted; no one could change it. Not even, I, for a long time, could change it. I flatly rejected any external idea that contradicted my beliefs about myself. I went all the way through high school knowing with absolute confidence that I was stupid and never saw any need to truly apply myself. Why should I, when I was incapable of learning?

Regardless of my own self-perceptions, I became first trumpet in the high-school band and was the best student in my drafting class. I still justified my stupidity. I thought, how smart do you have to be to blow on a horn and scratch straight lines on some paper?

Yet it was my drafting skills that later, much later, led me to Hampton University to study architecture. Looking back over my life now, it amazes me to see just how blind I was to my own abilities and how determined I was in defending my blindness.

I reached high school, reading on a marginal level, spelling atrociously, and being poor in math. All of this gave full support to the "I am stupid" belief. When someone would refer to me as one of the intelligent students, I could only think to myself, "I really have you fooled."

By the time I reached the age of twenty-eight, my mind was filled with so many distorted and confusing thoughts that I gave them a name, the "feces of the bull." The bull being the people I trusted for truth. I had been fed so much feces from those bulls; "feces of the bull" seemed quite an appropriate description for my confusion. Because of my confusion, I started having what are now known as panic attacks.

To give an example how a person's body reacts to fear, try imagining you being out in the woods hiking, and you came face to face upon a bear. Your conscious mind quickly reviews all that it has remembered about the face of this animal as you scream, "Bear!" Adrenaline secretes into your bloodstream; your heart pumps faster in preparation for your footwork. Your mind, heart, and feet are now ready to work in unison to take you in the opposite direction of that bear.

Bears have been clocked with the capability of running up to thirty-five miles per hour. Therefore, it is thought now that it may be best to just lie on the ground in the fetal position to protect your face and head by covering with your arms and playing dead.

My panic attacks occurred when my stored data could not

match, or provide a resolution for, the incoming data. My thoughts would become so conflicting with each other that I could no longer think straight. My mind would just start to freak out as in a dilemma, causing my brain to secrete adrenaline into my bloodstream.

Imagine yourself, instantly transported down into a Mexican town, you find you are being chased by several bulls down a small street. They are gaining on you, and as you strain to run faster, you feel a sharp prick in your buttock area. You quickly realize that you are being maneuvered out on the horn of a dilemma. You can now appreciate the fear in the dilemma.

I, on the other hand, found my attacks to be even more uncomfortable, especially when I had no scenario to view. It was like being on an alert, and yet, not at all sure why. I did not know if I should flee or fight. Even deciding whether to flee or fight was confusing. From what should I flee? With whom should I fight? The adrenaline flowing through my body could not burn up in my muscles because I was not running away or engaged in combat. The adrenaline was all going to my heart, working it overtime. With my heart getting that kind of abuse, it felt like I was having a heart attack.

So there I was experiencing such panic attacks and not knowing why. This led me to believe that perhaps my subconscious mind must then be a mechanism capable of thinking on its own. Nevertheless, *to think* would then mean an ongoing consciousness of its own at my subconscious level. Then my subconscious must be capable of examining data from my memory and coming to some conclusions behind the scene.

I then had to conclude that my subconscious mind roams unrestricted by rules and logic. How else could we have such dreams that, many times, do not make sense to our logical conscious mind?

It was clear to me that my subconscious, unrestricted by logic, was causing most of my problems. In trying to find a solution, it became necessary for me to clean out any feces from the bull I could recognize. I blamed no one for the lies instilled in me from childhood; I simply knew that I had to go back and rake through my mental pasture.

Having to go back and examine everything that was ever taught to me was a good thing, not only did I clear up the cognitive dissonance, I became a seeker of truth. Once again, I wanted to examine and know the truth about everything I believed.

# CHAPTER 3

## Purifying the Feces of the Bull

Not long after I realized that my pasture was full of feces that I decided to get some professional help with this purifying work. What really brought me to this crisis started when I was stationed in Italy from 1966 until 1969. During that time, I played first trumpet in the army band. I had also put together an off-duty, rock-and-roll or rhythm-and-blues band, which seemed very popular with the Italians at the time.

The guys in the band asked me to join them in the smoking of a joint one night after we had finished a gig. Being one who had been taught that smoking pot was wrong, I had difficulties with the idea. They finally convinced me, however, that the pot would help me free up my ad-lib trumpet solos.

The year was 1968, and smoking pot in the military had become almost a common thing. I am not saying that the military condoned it; I am only saying that it just happened a lot during that time. Well, I did enjoy that night smoking with the guys in the band. However, the next day, I quickly realized that playing uninhibited solos on my trumpet were not what I really wanted. What I really wanted was a better understanding of me and to know the truth.

I dissolved our off-duty band and started spending all of my extra time looking within myself. I started wondering how I could have become the first chair trumpet player in the army band, when many of the men competing against me for that position had held bachelor's degree in music. How could that be, with me being as stupid as I knew I was? Why was I always able to find a workable solution to most any problem or task the band had to deal with? Why was I known as "the man with the plan"?

The army band traveled as much as three times a week to promote a better Italian-American relationship. Because of this touring, we had to spend a great deal of time on coach buses. Onboard the bus, we were limited to what could be done during the rides. We mostly spent the travel time sleeping, playing cards, or reading. I choose the latter. For the first time in my life, I really started reading books. I read quite slowly at that time, but I was reading the books that interested me. Books on philosophy, physics, and some self-starter books like *Think and Grow Rich, Think Big.* Reading those books brought about ideas that conflicted with my established beliefs.

I remember my dad trying to teach me to read when I was in the second grade. He did not have to teach my older brother to read because my dad sent him to a private Catholic kindergarten for his fundamentals.

Dad started me out in the first grade. I was a good artist and the teacher's pet. Her name was Ms. Preston. Just after

Christmas, December 26, 1946, my dad got a job on the other side of Savannah. So, we moved from eight eleven West Thirty-eighth Street to Nineteen Pondus Avenue to be near his work. My older brother, who was in the third grade at the time, had to leave private school because there was no private school in that area for him. Therefore, he, too, entered public school. I should have been going into the second semester of the first grade, but the first grades were all filled.

Someone came up with the bright idea that I should go to the second grade, which was starting its second semester. This was a bad move in view of the fact that I had not gone to preschool or kindergarten.

My daddy took it upon himself to bring me up to the second-grade reading level by teaching me to read at home after his long, hard days at work. Every evening, when he came home from work, he would get a basin of water and put it on the floor. He would then take off his leather belt and start soaking it in the basin of water. Then he would sit on a low chair just about my little-butt level. I knew the drill. Drop my pants, turn my back to him, and start reading.

Reading became one of the most terrifying experiences of my life. I never knew when I was going to be struck. I was stricken even when he thought I said the wrong word, because he was not able to read very well himself. He would tell me that I was just too stupid to learn; knowing that I was his son just enforced that idea within me.

Years later, I would still have these overwhelming anxieties that would paralyze me whenever I had to read aloud. You could imagine how crazy I must have looked to people, sweating and looking out of the corner of my eye, as if I was waiting for someone to strike me. This is what I called cognitive dissonance. My mind would become flooded with the emotion of my past traumas. Cognitive dissonance would occur when I was not sure if I made the right decisions or not. This, then, would also cause an adrenaline rush, as I wanted to flee from this confusion. Such were the kind of things that drove me to visit the army psychiatrist.

At the time, I had been in the army seven years, and visiting the army psychiatrist was, by far, one of the best moves in my life. After a few weekly visits with this wonderful Dr. Nadine. I would like to say "thank you" to him, because Dr. Nadine treated me with respect and as if I was an equal. This was not common practice for an army officer to treat an enlisted man as an equal. He convinced me that he could help me. I remember him telling me that I reminded him of a good friend he once had in medical school. You would be amazed if you knew just how much a little offhand remark like that affected me—just to think that I could be like someone in medical school, someone respected and intelligent.

One of the first things the doctor had me do was to go over to the education center and take a battery of tests. I tested in math, English, science, and a controlled reasoning test, which was required that I be observe as I took it. I did dreadfully poor on all of the tests, with the exception of the controlled reasoning one. I only missed one question on that. When the proctor and the education director brought me in for the results of my evaluation, they first showed me the poor scores I made on the academics. I started feeling so humiliated and stupid. Of course, these were not new feelings to me, for I had long learned to handle them. I would curl my toes so tight in my boots that they would hurt. Then I would try to just melt out of existence.

The education director said, "You did quite poorly on these tests, but don't feel bad. Look at your reasoning scores."

I looked at a score of 99 percent. "Was this the test with all of the pictures?" I asked.

"Yes," he replied.

"So what does all of that mean?"

"It means that you are one of the brightest minds on the planet."

"You mean, I am not stupid?" I asked.

"Heavens, no. The only people who make these kinds of scores are your top judges, physicists, some CEOs, and writers." My eyes started filling with tears as he continued to speak. "Even your average school teacher only scores around 50 percent on this test."

I broke down with my head on the desk and cried like a little baby. Now I know this was not the way a twenty-eight-year-old soldier should act, crying like that. But what could I do when all of my life I was thoroughly convinced of my own stupidity? The female proctor gave me a box of tissue, and she and the director just gave me a few minutes to let it all out. When I finally raised my head to wipe my eyes, they too had tears in their eyes!

After we had composed ourselves, the director said to me, "Now, here is what you should do. Go back to high school in your off-duty time and reestablish your fundamentals."

"OK, I'm going to do that."

With the army band traveling throughout northern Italy three times a week on concerts, it was just not possible for me to start high school over again during that assignment. So, I got a library card and started checking out books to read on the bus.

I was so impressed with Socrates; his description of himself was that of a simple man, when in fact, he was one of the greatest philosophers who had ever lived. Plato, a follower of Socrates, became so outraged with Socrates' unnecessary death; he started a school of reasoning on Socrates' behalf.

One of the most life-changing books for me was *Think and Grow Rich*. I had no desire to become rich, but this book by Napoleon Hill really started making sense to me. It convinced me that I should receive an adequate compensation for my talent. Somehow I grew up feeling that I should get just enough to survive. This is another example of the kind of bull feces I had to clean up.

The feeling of worthlessness, I was taught, started to subside after I recognized this feeling as more "feces of the bull." Realizing that I could never forget my negative past experiences, I learned to apply a little technique which I called "bull feces purification" (BFP). This technique allowed me to bring to mind a negative set of thoughts, examine them, and attach something more positive and sensible to them.

It was necessary to attach a positive set of thoughts to the community of negative thoughts within, because our thoughts and experiences seek out like thought patterns that are already stored in our brain to attach itself to.

Once a thought finds a counterpart, that counterpart allows the thought to attach itself to it. The thought then becomes more like a next-door neighbor in its new house. Now these two neighbors work together in seeking more like neighbors (like thoughts). As they become a community of like thoughts, a combined charge starts to develop. These thought groups also have its passive thought counterparts in the area where sight, hearing, and feelings memories are stored also.

Whenever an alien thought is introduced, and there is not a counterpart to be found throughout any community, the thought is just rejected; no changes can be made.

There were just no like thoughts in my memory banks to permit me to store such a conflicting thought like me being intelligent. I habitually rejected my intelligence. I am tolerant with other bigoted people, because I was also prejudiced of myself. Therefore, it is understandable if some readers cannot find thoughts, which will embrace the connection idea. Sometimes new ideas take more than one generation to truly take hold.

So let us bring forth that community of thoughts which supports the "I am stupid" belief. I can embrace the feeling that went with it. I can experience the humiliation as well. Now that this community is lit up, I agree with all that I feel and accept it as a past truth. I am now in touch with and in harmony with myself. Just when I start sinking to a new low, I remind myself of that 99 percent score this stupid little guy made. The 99 percent thought is as if a banner scrolled out over the "I am stupid" community of thoughts.

Now, whenever something arouses that community of thoughts and causes them to light up in an attempt to take over my consciousness, the banner is also illuminated. Even if I am asleep and this community fires up in a dream, the banner illuminates as well, reminding that community of its true ability to overcome the past. This is a typical example of my "BFP" work.

# CHAPTER 4

## Quantum Physics?

In my search for truth, I found philosophy and science to be very interesting. But quantum physics has been by far the most intriguing study of all for me. The name quantum comes from the word "quantity," which means to quantify. Physicists call the small subatomic particles they can measure a quanta. Units of quanta are therefore the quantum particles. The ongoing studies now being conducted to understand the physical function of these subatomic particles are called quantum physics or quantum mechanics.

Because of the extensive use of mathematics physicists must use to communicate their ideas, they have all but lost their communication with the public, when in fact, physics is not above anyone's head, provided they have the interest. The extensive math had to evolve as a way of understanding and categorizing their findings. A standard method of calculation would have to evolve as a way for universal understanding. Like written music, anyone who can read music can replicate the sound the author had in mind.

Anyone understanding calculus can understand the mathematical calculations. It is all about learning an accepted set of rules.

The word "physics" should not make anyone nervous. Physics is nothing more than the derivative of physical, the activity or mechanics of a body and science. The study into the physical functioning or the mechanics of a science is therefore, phy+sic = physics.

Now that we are all comfortable with the title quantum physics, we are now ready to take the next step, and that is to become flexible in our beliefs. This is necessary, because the mechanics of quantum physics does not obey the classical laws of physics, at least not in the way we were taught to believe things should work.

I became comfortable with this radical idea when I looked up the word "belief" and found, just before it, the word "belie" which means false. I immediately recognized that some great mind from our past created the word "belief," so that we would understand that what we believe to be truth today could be a lie tomorrow. What this meant to me was that I should work with "belie" until (f) future findings come about.

If you can become comfortable with my belie(f) concept, you will not have any problems with your quantum connection. Know that everything changes, even what we believe at any given time is subject to change with future findings. Our universe is forever evolving with us as part of it all; we, as well as our thoughts, must be dynamic in its dimensions as well.

I see no reason, bar technology, for all the reverence over quantum physics when we, all of us, are quantum energy machines. Each time we bring our wishes to reality, we are wheeling and dealing as quantum mechanics and not even knowing it.

I have been able and am able to create my wishes over some period of time. This gives me time to reevaluate my wish by putting it through what I call my little "( )" test. Will this wish hurt the economy? Will it hurt anyone? Is this wish out of some primitive fear, anger, hate, envy, greed, or prejudice? If not, then so be it. I then fuel that wish with desires.

Desire is what let ( ) know how much I really want it. Incredible as it may seem, I always get these wishes. I have also put out wishes then realized later that the consequences may not be all good. Whenever that happens, I make adjustments before the wish has had enough time to fully materialize.

I have learned, however, that I can have instant gratification when I am in some prodigious situation. Some forty years ago, I was traveling across the country, and as I sped up my little black

Renault, in an attempt to move back over in front of a semitruck I just passed, I found myself skidding along the freeway sideways with the truck barreling down on me. Quickly, I oversteered to my left. Now I was skidding down the highway facing the opposite side of the freeway!

Now the night was cold with temperatures nearing twenty-five degrees, and the road was snow packed over ice, so being the intelligent guy that I had learned I was, I yelled out, "God, help meee!" Like magic, my arms steadied as I went into what felt like some kind of a slow-motion trance. It felt like someone else had taken control of me. Within the next right-left-three-two-one skids, I was back straight on the freeway. When I found myself back in control of my body and my car, I glanced in the rearview mirror, remembering that truck. To my surprise, that truck that was right behind me had backed off some five hundred feet or more.

It was not until then that I became frightened. So frightened, I could no longer press the accelerator without my feet trembling. I had to get off at the next exit and take a motel. You see, I felt that I was not the one who had pulled me out of that prodigious situation. That was when I learned that I had to be in a reciprocating way connected with something greater than my consciousness.

One thing I came to realize was that Jesus, like us, had the ability to manipulate quantum energy at his will. He demonstrated that his was the way that we should follow. However, if one is not taught the truth about his or her abilities, they could spend their lifetime floundering it.

What is being offered in this chapter is the chance to become free of fear, hate, jealousy, greed, prejudice, selfishness, poverty, and ill health. Name it and believe enough to rid yourself of it.

All of the above attributes are negative, yet such attributes are shared in some degree by all of us. Even after we have learned that these attributes are detrimental to ourselves as well as to our society, many of us still maintain them. So how does one detach oneself from those destructive attributes? One way is to decide what is to be gained by eliminating such an attribute. This will then facilitate in finding truth.

There are many truths to be known about many things. Then what is the truth? To start with, one should know that we live on a rock in space which we call earth. This rock is an ordinary mass orbiting an insignificant sun. This rock is not static, though we act as if it were; it is dynamic and in a state of constant flux as it yields to the laws of physics. It would appear that we as human beings are the product of the rock's flux and, like any other earthly by-product, are just another organism trying to find ways to stay clear of the rock's flux. We know that as tectonic plates shift, there will be earthquakes. As lava flow builds, there will be volcanoes. There will be, as some call it, destruction over the land. However the truth is, the earth is simply reacting to physical laws.

It troubles me to hear the media say things like "an act of God" or "the wrath of God." God did not do anything like that. What is happening to our planet happens to all planets as they evolve. The truth we must learn and accept is that physical laws, which are always enforced, govern the earth. We should learn all that we can about the forces involved, so that we can stay out of harm's way.

Another truth we should understand and accept is that we, like all other earthly organisms, are the by-products of our planet, as is our planet of our solar system, as is our solar system of our galaxy, as is our galaxy of the universe. We are the primary by-product to which reasoning has been bestowed. We are the thoughts and memories of the universe. We are, therefore, the eyes of ( ).

Because we are developing into the commissioners of ( ), we have been given our evolved brain which is capable of influencing the fundamental building block in the creation of all things. We must learn that we are cocreators of the universe. This may be a tall order for many of us to accept. That is okay; if that is your case, then accept it in small steps. No one wants to step out of his or her comfort zone too quickly. We are all afraid of the unknown. I suppose most of us would not have wanted to move out of the cave either. I am sure that it was only out of necessity that forced man out of his cave.

Changing the way we think is like moving out of our caves. Right or wrong, we have become quite comfortable with our old and proven way of thinking. It has been nice to draw on a set of clichés to take us through a particular situation—that is, birds of a feather flock together.

Do not think for one minute that as humans we're being led out of the caves; everyone just came out. You know there were those, keepers of the problem, who were too afraid and stayed with the comforting word of those who were also afraid. Well you know what? That was where they perished too.

Today, through quantum physics, we must be led out of our caveman mentality once more. Again, it is through necessity that we will be the driving force. I am sure many of the cave dwellers did not see the necessity when they had a nice rock perched high in one corner of the cave communities. Well, for those who have not noticed, this earthly rock is a dangerous place to bring up kids. It is just downright dangerous for humanity. Even though this rock gave rise to us, we must take deliberate steps to get off before our season is up.

If you have accepted the fact that the laws of physics have no consideration for us, then we must learn and respect its laws. Building a house on a fault line is idiotic. Would you do such a thing? Yet, look at San Francisco and so many other cities around the world. Now, how do you explain that?

My guess is that they were unaware when the cities first took root. But what about the people who keep flocking there after the faults have been well publicized? Perhaps, they went looking for and found a better life, becoming so comfortable that they just could not move out of their comfort zone. After all, we are but creatures of habit, and it does not take much to keep us there when someone else is there to speak some comforting words about the cave's benefits, even when it is doomed.

Another peculiar thing we hear people ask after some disaster, "How could ( ) let this happen?" Perhaps we should put it on ourselves for not learning and understanding the dynamics involved in the physical laws that govern our world. If your home slides down the hillside during a mudslide, ( ) is not to blame. Likewise, when hurricanes demolish our properties, again there was something we did not consider or know. We need to keep in mind that the earth is only reacting to the laws of the universe. No one is punishing us, not ( ) or anybody else. We must learn the laws and how to foresee the consequences.

Personally, in order for me to do this, I thought I had to learn as much as I could about our planet's ecological system, the biological system, as well as the sociological dynamics. Above all, I thought I needed to have a better understanding of quantum mechanics. The most important thing I learned, however, through the metaphysical training that I received was that we are all connected through our higher consciousness.

You do not need to be a rocket scientist, a quantum physicist, or anything else to be connected. That is because through human evolution we are already connected. Not knowing how to access your connection is like having millions in a bank and having no knowledge of the account.

Another very disturbing belief taught to me as truth was that I was born a sinner and that I had to be purified in the blood of Jesus. Being a very sensitive individual, I had some real problems with that. I could not see where I had anything to do with how I was brought into the world. What's more, I am quite grateful to my parents for the so-called sin they committed in getting me here. We are all brought here through the same birth canal as Jesus when he came through. So there is no need for apologies, no need to feel ashamed or dirty, no need therefore to ask for forgiveness. Jesus taught us that we should love one another and that is all that is required of us. Is it so difficult to see that if we just love one another everything else would fall into place?

Why try to make more out of it? Jesus did not want us to, and he never asked us to worship him. He wanted us to follow him. Follow him by loving each other. He did not die for mankind. He died because of the ignorance of mankind. The establishment felt threatened by him. That is why they killed him.

So let us quit trying to sugarcoat Jesus' death. You know, if they (the establishment of his time) were not so stupid and frightened, they could have had even more wealth and power by allowing Jesus to bring all people together in a loving society.

When I say the establishment, I am speaking of the religious leaders as well. Jesus was saying things that were not popular to the religious power base of the time. I gamble that the things I am saying now are not so unpopular in this time, because, unpopular or not, someone needs to say it, and say it now before we demolish ourselves. We must quicken our progress in coming together, so that we can move on, as a planet, while we still have time.

I am sure some frightened keeper of the problem is going to try to suppress this book through some form of discrediting and criticisms. However, just remember, his criticisms are only out of his fear. What fear? His fear that people may start thinking outside of his box. If I do not tell it now, someone else will have to try to tell it later, and it might be too late. The big problem with later is that we may not have enough time to change our ideological direction in time.

Jesus said, "These things you see me do, you will do, and more" (John 14:12). We will never do the "more" in which Jesus referred to, as long as we refuse to acknowledge the ability within us. Like Jesus, we all have the ability to manipulate quantum energy. What's more, we manipulate this energy now without even knowing it.

Many of us have been taught to seek outside of ourselves when we are in need. For example, "Pastor, please pray for my house and me." This type of thinking relinquishes your power to outside of you. It is better to say, "My house is blessed, and so be it." Both statements yield the same results. However, the latter keeps the power within you.

You may have noticed that your spiritual leader would always make sure that you believe in what you ask before praying for you. That is because they unconsciously know that you must manipulate the quantum energy required to bless your house as well. By believing in the minister, you then believe that it could and would get done through his help. Therefore, the energy required for the blessing of your house was instantly put into motion to form a new reality and direction over your house.

If we, as human beings, are to achieve the desires of our hearts, then we must first know that we have the ability within ourselves to achieve and create our desires. I have learned that it is unnecessary to go through a second or even a third party to get my wishes and blessings. ( ) has given me the power to bless me and my house. One must become aware of their quantum connection if they are to properly create and manipulate for themselves.

If every human being acknowledge that they are just one brain cell in ( )'s big old head, they would have then made the first step in their connection. How all this works is not so important as long as we know that it does work. It may take another century before we truly discover the physics involved in understanding ( ), if then. But fear not, that is what we are here for. How else are we to be the ears and eyes of ( ) if we cannot even gain an understanding of our own abilities? We are to understand our universe as best we can and pass it on. What will you pass on?

The quantum connection benefits us all. It means more business, more trade, more service, and more goods for all. It means a world market for all to share in. It means more peace and harmony. This is a win-win situation for all of humanity. Most of all, we will be moving in a direction that will develop us to fulfill our true purpose. Instead of making military hardware, those builders could build space ware as space becomes our economical catalyst. What will be so neat about a space-driven economy is that we can slow or increase our space output as needed to control inflation or other facets of world economics.

People like my dad use to say things like, "If ( ) wanted man to fly, he would have given him wings." Another good one was, "If ( ) had intended for the races to be together, he would not have separated them." Well, we are flying, and we have been migrating from Africa since the beginning. I hope everyone knows that by now. We can now see the implication that we are, in fact, all just one people. We are all just a bunch of migrating Africans.

So let us stop the madness and lift the whole planet up. ( )

cannot think clearly with deficient thoughts from us. There is so much more ( ) has for us to do. Now let each individual decide for him or herself that they will no longer be one of those who still make ( )'s head crazy.

Before we go any further, I want to ask that no one ever try to push the quantum connection on anyone. You should only expose them to this book and allow them to think for themselves. Most of all, do not try to make a religion out of the quantum connection. You would not make a religion out of the functioning of farming, would you? Yet farming requires a considerable amount of understanding to have success at it. We must see the quantum connection as farming; a considerable amount of understanding in the area of quantum mechanics must be had if we are to truly harvest our human crop. Understanding our quantum connection will unite the whole planet as we learn to capitalize and not demoralize, harmonize instead of brutalize.

# CHAPTER 5

## The Quantum Connection

The quantum connection simply means that we are all connected to the universe through quantum energy. This energy, when set into motion, is detectable as waves, which can be transformed into particles, particles that are the fundamental building blocks to what we perceive as reality, the foundation of all matter and all that matters.

Indulge me with your imagination as I take license in the postulation that our physical universe is but a sea of pulsating magnetic sources, observable through the potentiality of electromagnetic energy, which is only seeking its original state of rest before the big bang.

What does all of that mean to us as humans, and along with everything else in our universe? It simply means that everything came about through this "rest-seeking process." Energy cannot be destroyed or created by us; we can, however, alter its state. We have the ability, through our subconscious, to transform the wave energy into fundamental particles that make up our reality. This is the same process evolution takes eons to transform into products of raw materials.

If there is one thing we should be thankful for, it should be "the big black hole" (the primary black hole at the center of the universe). Because of its persistence in pulling everything back into it, the velocity in which things innately were hurdled out into the universe has resulted in what may appear to be a balance by resistance through orbiting.

As long as we—meaning everything in the universe—can maintain the proper velocity, we should be able to keep our orbiting balance and not fall as victims back into the black hole. Yet, even with the balance, everything is being pulled back into the black hole as the universe continues to be spreading apart. Why would this be?

The black hole, through electromagnetic waves, has a grip on the entire universe, and the more mass it crushes, the more electromagnetic force it gains due to the law of Conservation of Energy.

Not only is this universal electromagnetic wave force pulling us in, it is also through an invisible friction that decays the momentum of all orbits. This is really a good thing however, because without this electromagnetic wave field functioning, there can be no motivation for any creation within the universe; this concept is further explained throughout the book. Everything is therefore the result of the "big black hole's" pull.

Through this evolutionary pull, the milky way, solar system, planet, trees, water, air and all living things have evolved from this wave function to quantum particles (matter), to atoms, to molecules, to cells, etc.

It may come as a surprise to some to find out that we are "cocreators" in this process as well. Becoming aware of one's abilities to create or not to create can make a difference in their lives. However, I still wonder if it would matter in the universal scheme of things.

In view of the relative instant our consciousness has operated in this human form, the evolution of our consciousness will most likely look back on the human form we find ourselves in now, the same way it looks back upon the amoeba.

If we can manage to love one another and not destroy ourselves as we evolve our consciousness, we will evolve beyond the human form we now recognize as our identity. In trying to imagine such future consciousness, I am amazed to see how our social status has isolated us, even causing us to look down on others as if we have arrived. It would be better if we learn to embrace the idea that we, our consciousness, are still embryonic and have not arrived.

Formulating this Quantum Connection Theory has taken me a whole lifetime in figuring this much out, and still there is much more work to be done. I very innocently started this quest on a freeway in 1964, during my first three years in the army, while I was speeding on the autobahn near Nuremberg, Germany, trying to keep up with some young German teen in a Porsche. I could only watch the Porsche pull away from me as it moved back and forth through traffic from one lane to another. No matter how I tried, the openings just did not appear for me. This kid was moving about twenty miles per hour faster than the two lanes of traffic, yet he never had to slow down.

I, on the other hand, was stuck with the meager ninety-five-miles-per-hour speed of the moving traffic as I watched those spaces open just enough for the teen when he needed them. What's more, the spaces closed as soon as he moved through them.

I could see that something else was going on. I knew that this was not just luck. I even considered that the Germans might have so much respect for a Porsche that they are willing to slow down to allow the Porsche to move over into their lane. I had to reject that idea because I did not see anyone's brake light go on. What is more, things were moving much too fluid for someone to react that fast. It was like that kid knew that a hole, in the next lane, would appear when he got close on the bumper of the car in front of him. It did not look like the kid just hoped the holes would appear; he drove as if he knew they would appear.

Driving around on the autobahn at one hundred plus miles an hour was, however, a bit too risky for me to experiment around with, so I decided to try something less risky. I started believing in a parking space up front in a mall whenever I had to park. I would think, "I want a parking space up front." At first, I had to drive around the lot two or three times before I finally got one up front.

What I later understood was, as I drove around looking for that space up front, that my desire for the parking space heightened. It was true desire that started the magic. (Well, it seemed like magic at the time.) I would drive to the front expecting my space, and the space was there on my first pass.

I developed the habit of expecting my parking space to be up front. Even when the parking lot was very full, someone's reverse light would come on, and all I had to do was stop and let them back out. This worked all of the time except when I had an observer to whom I had bragged about this ability.

Reason: bragging has the tendency of putting others on the offensive. This offensiveness usually manifests itself through a disbelief in the endeavor. Now instead of having someone believing with me, I inadvertently created an opposing set of energy coming from that person to whom I had bragged. The person may not be deliberately trying to oppose me; it is just that they may not be able to see themselves achieving such a quest, so how could I? They then expect a different outcome than mine. I then end up with no front parking space. Lesson: braggers do not know; knowers do not brag.

Through simple experiments and meditation, I started to learn that my thoughts have an effect on things. At first, I thought that I was directly influencing the thoughts of others. Something like, my will over theirs. However, this was not at all the case.

I later discovered that our thoughts and wishes materialize through our desires. Thoughts and wishes are just like miniprograms stored as electrical impulse in a chemical memory bank. Each time we go back and think about a particular wish, that thought is also sent out into the universe as a wave transmission. It is in the universal consciousness that such a wave will line up with like waves.

Thought wave, once in the universal consciousness, cloaks itself in the ubiquitous electromagnetism, which can then be transformed into its particle equivalence, a quantum particle. So how does one get from wave to a molecule? What mechanism can be used? How should one bring about the physical particles from its wave component?

Physicists are still pondering over that phenomenon, but it does not have to bog us down. What is important for us to understand is that we have the ability to facilitate the transformation. The mechanism should now become obvious.

The mechanism is desire. Once the waves have been sent out into the universe as a thought, it is essential that one holds fast to their desire of that thought. Desire is what brings your thought back to you in the particle form. When we create a real desire for that thought or wish, the thought does not just float around in the universal consciousness soup; well, it will if desirability is not chasing it.

Desire can be viewed as a big soupspoon reaching into a bowl and lifting out the part you want, that swig of soup you are most familiar with, your wishes. As you place it into your mouth, it turns into your reality.

Desire is the transformer as well as the magnet that brings about the physical reality of one's dreams. Desire transforms the wave into the particle. Could it be that when like waves line up, they crystallize as a particle? Again, the physics of this process is yet not understood, so why not work with my idea about desire for now?

We are all cocreators; the universal consciousness, influenced by all of our desires, accommodates us all. All that one needs to do is first be aware that they are cocreators, when they truly desire and believe without a doubt for what they wish. You do not need to make it happen; simply desire it and let it happen.

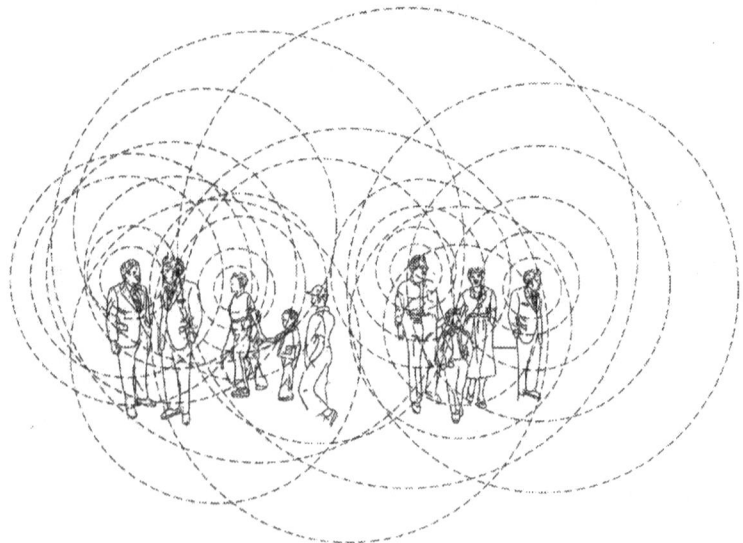

The world does not revolve around you ... your whole universe does.

What we feel, think, and desire influences the universal reality. We are a part of an unfolding whole, where any reality is achievable and where all things are possible.

Now that we understand that notion, we must also be aware that fear is a negative desire. Therefore, focus on what you want, not on what you do not want.

I was once one of those people who thought that you must work hard to get anything in life. That was just some more feces I had to purify from my mind. I had to figure out that my beliefs were governing my reality.

There are many ways that I could explain the quantum connection. What's more, there can be no best way to explain it, because we all have been taught many different beliefs. So whatever higher consciousness you believe in, you may try substituting the name of your higher consciousness whenever you see "quantum connection." You will then find that we are all talking about the same thing. What I have done is simply put somewhat of a different spin on it, a spin that allows us to think just a little outside of the box, a spin that allows us to live in harmony.

To ensure against any psychological dissonance, I suggest, once again, inserting what you hold as "Most High" into this ( ).

Imagine for a brief time that ( ) seeks to understand itself by creating us as its instrument; why not? Keep in mind that it is ( ), and ( ) can choose any method it wishes to study itself. So, if ( ) has chosen this way, through us, to become fully aware of itself and all that is around and about itself, then who are you or I to resist? Who are any of us to resist ( )'s will?

Now try to visualize and believe that every human being is like one brain cell in ( )'s big brain. Let us see this big brain as relying on each of its cells, us, to record and understand itself. Imagine what your head would be like if your brain cells were constantly trying to harm other cells in your brain. Cells forming groups of cells to promote their own self-interest, attacking and killing off each other. As new cells are born, they too are taught to hate other brain cells; for generation after generation, they never decide that they should be pulling together as one group.

If this was the head of someone you knew, would you say that this person were a troubled individual who needed to find some better approach for getting its mental house in order before they self-destruct?

What could you do to help fix such a person's head? How would you stop the craziness inside their head? You could not just sit there and feel sorry for them as their brain moves closer to self-destruction each day. What should you do? What could you do? Perhaps you could somehow just wash everything away and start over.

With such conflicting thoughts coming from the brain cells, and with each of these cells having a strong drive for self-preservation, a form of brain cancer that will ravish its own resources could only be the resulting outcome, wouldn't you say?

Recognizing this cancerous circumstance, it would appear that the most likely thing to do would be to create an antidote in the form of some new thoughts that could be embraced by all of the conflicting brain cells—thoughts that can attach themselves onto the confusing cells and bring about trust.

Stop, take some time, if you have not already done so, to truly imagine this as being your brain or of someone you know, because only you can do anything about the cancerous cells being mobilized throughout the head of this person. Do not think of this as just some futile game in imagination and which you do not have to come up with a solution.

The reason I wanted you to stop and really think about solving this imaginary brain cell problem is because this is someone you know. This is the head of ( ), which we are all a part of. ( ) seeks understanding of itself through the creation of us. ( ) is now trying to get its head in order. ( ) has the ability to create and give us, itself, all that we can ever desire. Why not? After all, it is itself. Each of us is like one brain cell, stored on earth, for ( )'s understanding and pleasure.

As you accept this scenario as in the head of ( ), recall our earlier lesson that thoughts cannot be totally alien if they are to be accepted in other communities of thoughts. As illustrated from my past, foreign thoughts are rejected. However, rejected or not, the craziness will, and must, stop.

As a solution, we could send in some refreshing new ideas that we feel can entertain and embrace the beliefs already established in the brain cells.

But wait, wasn't that what ( ) did a couple of millenniums ago, when ( ) decided that the brain cells needed to learn to love one another? An attempt was made to teach us to love each other: love your neighboring cell. However, at the time, such a thought was alien to most of the body of brain cells. In fact, when that love message came to earth by a messenger, that love message was so alien and threatening to those in charge cells; they killed the messenger. We have now assimilated enough of the love idea that we can admit that the message, as well as the messenger, was then misunderstood.

Even though the messenger was silenced, the love idea he brought, somehow, took root and stayed around and has now developed somewhat; we still have ways to go to identify with each other.

Keep in mind that ( ), like us, seeks understanding of itself, then recognized that before neighborly love can be thought to its brain cells, it is necessary to first make each cell recognize that it is part of its whole.

Now, a new attempt is being made to show each of us that we are all a part of ( )'s brain. We are all important to ( )'s body. ( ) needs all of our diversities and all of our interesting thought input. If we can be receptive enough to understand this, the head could finally be in order to do the things it wants to do.

One may think, well if it is ( ), why doesn't ( ) just take over everyone's will and make him or her do whatever ( ) wishes? Simply stated, it would be counterproductive. It would be like someone taking control of the Internet for their selfish desires and fear. This would surreally hamper creativity when it is known that a hampered mind profits little. With each cell left free to think creatively for itself, ( ) also is provided with the best brain for discoveries.

The love idea is no longer an alien idea to our communities of cells now. So this idea that "you are one cell in ( )'s big old brain" should now be able to attach itself to the "love one another" idea. The sooner we get it, the better. This is the "we are one" idea. If you got it, then by letting it take root in your heart this time is all you need to do. This idea can and will surely enhance all of humanity; what's more, it does not and will not take anything away from you.

The fact that you are reading my thoughts should prove to you that I love you. Yes, you may be a brain cell that I do not know, but I know that I love you because we are one. We are one in ( )'s big old head; what's more, there is nothing you can do to change that. So there, you might as well show me some love too. So come on, give you and me a smile!

To love all the other brain cells does not mean that one cannot promote your own selfish desires. What it means is that your selfish desires will not harm others. In fact, when your desire is governed by the love idea, everyone somehow benefits in the end.

When I project my desires to understand something through ( ), and that is simply by thinking and meditating, ( ) reflects back to me an insight which I chose to view. If you are not doing this, you really should be—do not miss out. Good things can happen to you. Things like, extended youth, good health, happiness, peace of mind, and, most of all, your justification of success. Not the success someone else wants to justify for you.

Most people are aware that there is some kind of ( ). If you are not aware, it could be because you may have been referring to it as just the ability within yourself. Well, that too can work, because in the final analysis, within all of us is an ability. However, this type of thinking usually causes one to believe that they are not connected to the whole, when it is essential that we all know that we are one.

Another problem that does sometimes occur with that, "me, ability thinking," is an attitude of "I got mine, and it's up to you to get yours." Often, this attitude is accompanied by an extraordinary ruthless selfishness. The amazing thing about those types of cells "getting theirs" is, after they have spent a season or so embracing their wealth with other like cells, they find that they are still not happy.

They are not happy; they know that something is just not right. Well, of course not, it is because they do not understand their connection to ( ). Do not misunderstand me; there is nothing wrong with accumulating a stupendous amount of wealth when you have provided a service to humanity. Wealth is ( )'s way of rewarding you. But for one to steal, with no regard for another struggling cell that may not share your advantage point or that may not have perfected your survival skills, is a discredit to themselves, to humanity, and, most of all, to ( ).

They are a discredit to themselves, because they must now turn to alcohol or/and drugs to attain a euphoric happiness. Really, drugs further the disconnection of them from their consciousness. Being disconnected from one's consciousness is to be disconnected from ( ), which must be the ultimate discredit to oneself.

I also said that they were a discredit to humanity, because thieves can only promote more thief consciousness into others. What's more, these ruthless thieves steal from society then sit smugly on the sideline as the rest of the productive cells seek to follow ( )'s will by obeying their ( )-given curiosity and what they know to be right.

If we create harmony among all the other cells, our human species can and will venture into the outer spaces at a much faster pace. We have the free will to go to the outer space or just sit smugly on the sideline and space out.

Earth is but our birthing place; we should not become too comfortable on this rock. Earth has been a safe outpost, away from the bother and the development and consciousness of other civilizations. It is now time for us to start realizing that we have a much higher purpose. We also genuinely need to start thinking about, "How shall we get off this rock?" Better sooner than later.

We need to better understand who and what we are within the universe. How do we really fit into what we see as real? How do we know that what we have been taught to believe as the will of ( ) is nothing more than the will of those with the control?

Looking back at history, one can see the many civilizations which rose up with their "as we can now see them" strange, unworkable religions. Any religion or beliefs that advocate the harming of other human beings will not, and cannot, last. In the final analysis, such societies have perished because of their lack of love and identity with each other.

If humans cannot grasp this notion, and grasp it soon, we will once more be doomed to start all over again. This time however, we can expect our surmise in the form of a nuclear winter created as the results of our opposing religious beliefs.

If you find yourself in a belief system, which advocates the harming of other human beings, then you may want to spend some time thinking about the inspiration of that author. The fact that we are so violent to one another proves that there are grave faults in what we have been taught to believe. One could say, "It's just the nature of man," negative. Our nature is capable of all things, including love for each other.

Better yet, we need to quit wasting time. Try looking up in the sky, on a starlit night, at the many movements. Is it so difficult to realize that something very huge from outer space could enter our atmosphere and ignite into a fireball, wiping us all out?

What should be realized is that we are not sent into existence to triumph over others. We all have a purpose that is important to ( ). Many of us are in existence to create better brain cells by teaching school, cooking food, and servicing the general welfare of the body. Such tasks are all vital to the whole body of ( ). All of our brains are needed to satisfy the curiosities and the enlightenment of ( ).

( ) wants and needs all of our inputs. So if one cell is attempting to have ( ) resent another part of its self, then that cell has some reevaluating to do. The thoughts we are sending out to ( ) should be thoughts that magnify its body.

If any cell senses some resentfulness in this message, it is understandable. But let them try not to judge before asking ( ) for understanding through their personal beliefs. ( ) will delight in just knowing that you are committed to understanding. We are the reality of ( ).

The name, or the way, in which we refer to and communicate to ( ) does not really matter. What does matter is that we all come up with a way to refer and communicate. ( ) wants and needs us to get this message, DO NOT FIGHT AND HARM ANOTHER HUMAN BEING. Fighting in the body is enough to drive even ( ) crazy. What we should keep in mind is that the earth has been cleansed before and can be cleansed again, due to the nature of our universe's evolution. I am sure none of us really wants that cleansing before we can make our transformation. Fighting among ourselves is but a waste of our precious time on earth.

An important thing that I have come to accept about ( ) is that the physics that is required for our true understanding of ( ) is not, has not, and may not be discovered by us for a long time. But even so, we should stay alert; someone among us may be enlightened in a flash. If it is truly ( )'s wish that we understand for its own purpose, then we should be studying and meditating for enlightenment.

People have asked me, "Isn't it enough just knowing that ( ) is?" For me, it is not because I have this spirit to discover and understand; it was not put in me to sit idle. This spirit has a purpose, and this purpose is for ( )'s understanding. So let us not be afraid to ask the difficult questions like "How does it all work?" and "What is the physics involved?" ( ) obviously wants us to ask these hard questions. Thanks to people like Albert Einstein and so many others before and after him, we are now beginning to make some sense of our universe.

This quantum connection idea should take hold, because it benefits us all. It means more business for all, more trade, more service and goods for all. It means a world market for all to share in with more peace and harmony. This is a win-win situation for all humanity. Most of all, we will be moving in a direction that will develop us to filling the true purpose of our existence.

We will never move to the next step as long as we continue fighting each other. So it is now up to us; only we can stop the madness and lift the whole planet up. ( ) cannot think clearly with deficient and conflicting thoughts from us. Through my mediation,

I have learned that ( ) has so much more for us to do. So, as I have decided, and so can you, not to be one of those cells that are making ( )'s head crazy.

# CHAPTER 6

## Manipulating Quantum Energy

In the last chapter, I postulated that, over eons, the universe has created the stars, planets, and all that we see through a process based on electromagnetism. If the big bang theory is correct, as scientific evidence points to, then as the universe expands, everything is just trying to find its state of balance and rest. The original momentum the big bang caused, even though deteriorating, is still influencing everything in the universe.

The universe may be quite simple to understand, as we gain a full understanding of electromagnetism. It was thought that once the universe reaches the extent of its expansion, it will begin its return back to its birth. If this was the case, it will return for the very same reason it is coming to a halt. There is a tugging through an electromagnetic friction-like force. This force is stopping the velocity of the initial big bang's thrust. If there were not such a force stopping this outward momentum, our universe would go on expanding forever.

The universe is expanding while being called back to its origins by the process of electromagnetism. Let's start with what we are most familiar with, the illusion we call gravity; what we call gravity is nothing more than atoms in a body, seeking balance between positive and poles of negativity, resulting from the big black hole.

No matter how high you may jump, you quickly find yourself plummeting right back down to the ground. What you witness was the functioning of electromagnetism. The ground with its greater mass has the greater electromagnetic field. Without this electromagnetic force, we would not have stopped when we jumped.

Fortunately, for us, this field has an insatiable appetite, and the more bodies the field captures, the greater its mass becomes; the greater its mass, the greater its appetite becomes as well. Mass is not what this force is after; however, it's out to capture the atomic essences of the mass, its electrical charges.

If not for the electromagnetism, we could walk through walls or any other object, including each other. Atoms are the building blocks for all things; without electromagnetism, atoms would just fall wildly onto and through themselves.

This line of thinking would seem to present an oxymoron. Without the electromagnetism, nothing would be in existence in the first place. Yet, the primary purpose for this ubiquitous field is to recapture and digest itself into the black hole. This would make the big black hole the *creator* of all things. This is what gives all things its purpose in being.

Finally, if this is the creator, have we located ( )? But wait, why would ( ) be so overwhelmingly big and black? If you are one who has been taught to attach negative feeling to what is "big and black," this could be the perceptional paradox that will or will not allow some of us to pass "go" in this book. Simply because anyone who is still sensing the need to ask this SAQ will need to stop here and ask themselves, "Does ( ) have that prerogative or not?" On the other hand, one could convince themselves that ( ) and the black hole have nothing to do with each other.

**A Childhood Photo of Chaz**

Now, now, do not become upset with me; remember, I am just that little Georgia cracker boy still trying to answer his SAQ. I said to everyone earlier, we must be willing to relinquish our preconceived notion if we are to hook up with our quantum connection. In other words, be willing to look beyond the conflicting thoughts in our teaching.

The solar systems along with their planets and moons would not have developed if not for this process. There is nothing in the entire universe that did not come about without this electromagnetism process. We, along with our thoughts, resulted through this process as well.

Gaining a full understanding of this function will afford humanity the most practical solution to all of its needs.

Simply stated, the earth wants to capture the moon while the sun wants to capture the earth. Bigger suns pull on our sun, and all suns are being pulled to the center of our galaxy, while our galaxy, along with all the other galaxies, is being captured by the universal big black hole. This big black hole is the creator, ( ).

What I have realized from this entire body-capturing struggle is that with the universe being full of these electromagnetic waves, we could, and should, be learning to surf that wave as a transportation mode. The universe must be a sea of electromagnetic waves; how else could the black hole be able to reel itself back in?

Well now, if we, as well as all other bodies in the universe, are the results of this electromagnetic wave process, the same must be true for our thoughts as well. Think about that thought.

Before you start your deliberate attempts in the manipulation of quantum energy, you should and must examine yourself to make sure your desires are pure and good for you as well as for humanity. As you buy into this whole quantum connection idea, you will have to buy into the fullness of the ideas in the next chapters as well.

As for your prayers, meditation, or however you have been taught to connect to ( ), an efficient way I have found to send out and receive what's desired is to, first, be acceptable to the love message. However, if you are still not quite able to love all of the other brain cells, your human counterparts, then see if you can simply make a conscious decision not to do anything that could or would cause malice to anyone. That is an acceptable start.

Next, stop any negative communities of thought groups you can identify within yourself. Use some of my techniques, or create some of your own, because knowingly or unknowingly, we all tap in and manipulate quantum energy. We are simply now going to become fully aware of what we are doing. After all, ( ) stores and remembers all of our thoughts. Just like our own brain, we store memories of our experiences for future reference.

Here is what happens when we have some desire. ( ) compares and locates other human beings, its brain cells, which have an accommodating desire, bringing them together through an alteration in reality. Remember, like thoughts can only gravitate to like thoughts. The same is true for like desires. This altered reality or future is usually a win-win situation for those involved. After all, it is what everyone wants.

It is said that people like Nostradamus and Jeanne Dixon shared the ability to see into the future. I think that people who display this unusual talent are really viewing the results of our collective consciousness as a future reality.

In other words, it is yours as well as my thoughts that will create any future reality for humanity. Be it good or bad, it's what we think that will govern what will come about in our future. What an awesome realization! Our past philosophies have landed us into today's crisis.

If we continue to think selfishly and with hateful thoughts, there is no need to hope for the best; such thoughts can only align with like thoughts. As these thoughts reach a critical mass, the realization of the impending adversity becomes apparent.

On the other hand, we could start a deliberate campaign that would facilitate loving, caring thoughts that will bring about a secure future for all humanity. Through the understanding and practicing of our "quantum connection," harmony is at hand.

For example, when someone prays they are asking ( ) to facilitate in their needs, when actually, it is only through the true desire of the unambiguous wish that motivates the wave function into the particles of that desired reality.

It must have been much easier for religious framers to teach that it is the prayer that brought about the wish rather than attempting the explanation in the extravagance of such a quantum event.

Learning to desire and then believing for what we want puts the power to create in our hands. Asking diminishes one's ability to help themselves. That should have never been the case when we are all cocreators.

Each of us shares in the spirit of ( ). It is this universal spirit that grants us our ability to manipulate the wave function into what we desire. Wouldn't it be more effective if we came to the realization that we are all ambassadors of ( ), and that in order to be an effective ambassador one is expected to think as well as create for themselves?

When we know that we are cocreators with ( ), we have no need to distract ( ) with trivial needs. ( ) is busy doing bigger things like making stars and planets, not to mention all the work going on inside the big black hole.

That does not mean that ( ) does not have time for us. On the contrary, ( ) loves and cares for us so much that we have been given the mental ability to cocreate our needs.

One would not ask the CEO of the corporation for their office supplies, when the CEO has already established an expense account on their behalf—certainly not. One only needs to know how to access their expense account. Accessing the account is what the Quantum Connection Theory teaches.

The quantum connection works like a perfect democracy. One vote per person—it does not matter if they are poor or rich, living here or there, educated or not. They get one vote, and that one vote is their thought input.

But, if they give up their mental democracy by allowing the ideology of any group that promotes hate, to influence their quantum input, then they become nothing more than a quantum energy generator for the ignorance of a totalitarian mental-state group. Such mental-state groups will only facilitate its own demise. The reality of World War II, for example, was the manifesting of this process.

My dad was a good example of one who manipulated quantum energy unknowingly in a negative way. Still I was able to turn his negative input into a win-win situation. See if you can find the win-win in this next circumstances.

My dad would take every opportunity he could, to tell my older brother things like, "Son, you are going to have something when you grow up." What this did was convince my brother to expect wealth when he grew up.

Contrary to what he programmed my brother, Dad took every opportunity to tell me things like

"Boy, you ain't never gonna have anything with your stupid self." Here, you can see quantum energy being put to work in twofold. First, he sends a positive set of energy for son number one and a negative set for son number two. As was earlier explained, ( ) does not evaluate our expressed energy waves. These expressed waves are merely accepted and vibrated into their particle components for the building blocks of that reality.

As my brother and I received Dad's productions, we unknowingly accepted his words as truth. After all, Daddy was "all knowing" to us at that time. With such thoughts locked in our subconscious mind, we certainly sent those messages into the universe ourselves. We unknowingly sent quantum energy from a belief that was seeded by Dad.

It should be obvious how our lives turned out. My brother, with only a high-school education, became a manager in a Fortune 500 corporation, which I will not identify. Even if or when his lack of education became apparent to his coworkers, the corporation would act for him. In fact, adversity usually turned into a promotion or some kind of upgrade for him. I am not going to share his business with you, but I can tell you this: before he retired, the corporation provided a chauffeur-driven limousine which picked him up, to and from work. Why? It was nothing more than his strong feeling of deserving. "I am supposed to have plenty, now that I am grown up." "I am supposed to be successful."

Last year, I took the opportunity to visit one of my old high-school friends (Arthur Demerit) down in Ft. Lauderdale, Florida. His class of 1957 was having its forty-fifth class reunion. This was not my class; I was in the class of 1958. My older brother was in the class of 1956. We all attend Booker T. Washington, "a separate but equal" high school down in Miami, Florida.

I decided that this would be a good time to take a little break from the writing of this book, visit some old classmates, and see just what their lifetime had dealt them. Little did I know, I would be getting material for my book.

Of the many successful people I met from that class (and they were many), some were doctors, lawyers, a judge, a publisher of a Chicago newspaper, a high-school principal, and some teachers. However, only one earned an honorable mention in my book, Mr. John Glover.

I choose John because his life's success exemplifies the material in this manuscript. Here is a part of an article, which I took from their class reunion information book. This newspaper article was published in September 5, 1982, by the *Miami Herald*, long before John had retired.

The headline read, "Miami-born man earns top FBI post ever by a black." FBI Director William Webster has named John Glove, head of the agency's Atlanta field office, as an assistant director in charge of the FBI Inspection Division. The move makes the Miami-born man the highest-ranking black in the bureau history. It was the fifth time in Glover's sixteen-year FBI career that he has been made the top black in an FBI division.

I personally recall John as a star high-school football player. As I was in the band, I witnessed him making many touchdowns for our school. He was not only an outstanding football player; John was also captain of his team and class president. Here was a youngster who has learned to be a leader from childhood.

John was always awed with a show called "The FBI." His childhood dream was to be an FBI agent when he grew up. If you are old enough to remember what it was really like during the 1950s, you can agree that, in the 1950s, there was no way Mr.

Herbert Hoover would have put any blacks on the FBI's staff. Most people knew how Mr. Hoover felt about blacks. Nevertheless, look at what can and did happen when you do not spend time focusing on the circumstances but, rather, just on the desire.

As John attended Florida A & M University on his football scholarship, he learned the reality of Mr. Hoover. Naturally, John started trying to think in a more realistic way by thinking he could become a pro-ball player. Yet, remarkably a set of events occurred which sabotaged that idea forever. Seeing his dreams for becoming a pro-ball player shattered, John decided to become a teacher.

A very curious thing happens when one has spent a lot of time dreaming a dream with desire. Even when the dream changes, it may be too late; the universe has already started crystallizing the waves into particles which will bring about the reality of that dream. One might as well be there to receive it when it comes.

Knowing John's background as a leader and an achiever of excellence, there was no way John could be content just teaching school. It is my guess that as John settled down into his job as a high-school teacher, it was not long before he became bored and unfulfilled. What we all do when we are not content with our lives, we go back to our first love. In the case of John and the FBI, rekindling that childhood desire was all it took for the universe to now usher in a shift in reality, a reality that started its crystallization and then had to sit dormant for years when John abandon it.

Little did John know, the universe had started crystallizing his wishes before he had even finished high school. Although many realities would shift to accommodate John's wishes, the preparation for the shifting in reality had already started. When the universe has accepted one's wants, quantum energy immediately starts its manipulation to bring forth the desired reality. Even when the sender tries to change their mind, there is little chance of turning back. The universe has already started setting into motion a reality that accommodates the desire.

Returning to that childhood dream once more with new desire, the FBI found themselves down in Miami looking for "him." The FBI knew exactly the kind of man they wanted, and John was that man. John had worked his quantum connection without ever knowing it.

One thing I should add however, if you are going to set dreams into motion through desire, try to dream the entire dream. For if you do not dream the entire dream, your dream will be filled in by the desires of someone else's dreams, because all realities work in conjunction with each other.

Because all realities work in conjunction with others, you may not like all the fill-ins that come with your dream if you did not fill in most of the blanks yourself. If this looks like I am suggesting that the universe can and will reorganize the whole reality just to accommodate your desires, you can bet that I am. What's more, ( ) is relying on us to create our realities. That is why all of us must have freedom of will.

We make the reality for ourselves as well as for ( )'s enlightenment. That makes each of us important to ( ). That is why it is so very important that we allow everyone the freedom to think and dream for themselves.

What we must understand is that there is no set reality. The universe is capable of accommodating an infinite amount of realities all at once. What's more, they will all fit together like a glove. In other words, your reality accommodates mine. However, for those of you who refuse to dream, your reality becomes whatever the dreamers' see fit for you in assisting them.

I am sure that you will agree that Mr. John Glover's story is nothing short of remarkable. One might think that he overcame what was, at first, out of the question. However, in truth, if you dream the dream with fuel of desire, it would be remarkably strange if it did not materialize.

We do not need to understand the quantum mechanics involved in the bringing forth of the dreams. All we really need to know at this time is simply to dream the dream with desire and know that the universe will manifest it.

I have been taking every opportunity I get to teach children that if they are going to dream, dream with desire and stay steadfast, so that when the dream comes, they will be there to receive it. If one is wishy-washy and always changing their mind, they will not be there when it comes. It will just fall into the lap of someone else who can fit the waves.

Let us get back to the success of John and my brother. With so much positive quantum energy forever vibrating from the minds of John and my brother into the universe, such thoughts and feelings like, "I am a leader," "I know that I deserve success," "I accept success," "I deserve, I deserve"—they could only receive success.

Believing in something is like staying tuned to a channel, a frequency, a wavelength of the dream you desired. Allowing someone to discourage you from your dream is like turning off the channel, leaving no way for one's dream to reach them, which was not what people like my brother and John did. They held on to, or came back to, their beliefs and dreams.

Any would-be challenger of these men would surely receive the quantum energy, which they vibrated into the universe. Challengers would find themselves conforming to the strong positive belief these men generated about themselves. Even their superiors would start feeling that here is a man that deserves to be elevated.

I have found that people with only a positive deserving belief about themselves—and sometimes with little or no talent even—can quickly learn how to get those with the talent to get the work done for them. We have all seen a boss take the credit for other's work. But that's another story.

In no way am I suggesting that this was the case in the life of these two men. I only want to take this opportunity to show everyone that all they need is the will and desire as they work with their quantum connection. When you ask for whatever it is you wish, with desire, it can only come to pass if you hold on to your dream.

Earlier I talked about how my father taught me. By the time I was in the third grade, I knew without a shadow of doubt that I was stupid and would never amount to anything. As you may have suspected, I never tried to learn my schoolwork. After all, I knew that I was too stupid to learn. I was also convinced that I did not deserve anything.

So I spent a great deal of my life doing creative things to help the other deserving people get what they wanted out of life. Well, why not, I dare not dream for myself. I became quite good at just desiring for whomever I was around, well, if I liked them and understood their dream. I would wish their dream with them. I always wanted to see them get what they wanted. My childhood teachings left me feeling undeserving, so how could I dare desire something for such an undeserving person as myself? ( ) gave me just what I asked for: to help others get their wishes. ( ) does not discriminate our wishes. ( ) merely accommodates the quantum energy by transforming them into practical possible realities.

As early as thirteen years of age, I can recall the first and most deserving man I ever met. He had such a strong sense of deserving. It was at my first after-school job. My brother and I worked at his restaurant at 550 Brickell Avenue in the downtown Miami area. The year was 1952. I was in the seventh grade and worked from 5 until 8 p.m. each day.

My boss had me dressed up in a white chief's suit with a tall white cook's hat. He gave me a dinner bell to ring for three hours. This was his way of stimulating business. Well, let me tell you, it worked. Folks would come from all around town to see "Dinner Bell Charlie" as advertised in the *Miami Herald* newspaper.

What I was able to do so creatively for my boss was to ring and flip that dinner bell rhythmically, almost nonstop for the three hours; I would switch from one hand to the other when one arm got tired. I would flip it over to the other hand without missing a beat. My boss was quite conscious of his profit, and I wanted to see him get the profit and success he deserved.

Recently while shopping for some parts for my sailboat, I saw a bell with a handle that looked like that old dinner bell. I had to go over and get my hands on it. To my amazement, I still had that rhythm after fifty years. The rhythm was in two-fourth time, with 2 sixteenth beats, a sixteenth rest, then 3 eighth beats, then flip within the last eighth beat, repeat over and over.

After a couple of years, of packing the people in the restaurant, my boss had made enough money to do what he really thought could be more profitable. He sold the restaurant and combined his cash with a friend of his whose name was Dave. I think that Dave was an accountant. I never got to know Dave.

I see nothing wrong with someone like me using their creativity to help someone else get what they want in life. Especially when I did not know, myself, what I want or could have in life. However, Jim and Dave did know what they wanted, and they founded Burger King.

James McLamore was the primary owner and was unwavering in his desire. As I stated earlier, if you don't dream, you will only serve the dreamers who know what they want. Mr. Mack could not take me with him when he sold the restaurant. I was part of the sale. I had to stay and ring that bell another two years for the new owner, Mr. William Billarket. I called him Mr. B. Mr. McLamore did ask me to work with him again in 1958 to help service their franchisee, which they had acquired in Florida at that time.

That was just after I had finished high school. I agreed to work with him again to make enough money to get away from my dad. It was more important to me that I break the grip my dad had on me than to stay in Miami under his influence. I recall Mr. Mack telling me that for eighteen thousand dollars, he would let me have a franchise in the "colored section" of town. In those days, that was like saying, "For eighteen million dollars, I would let you have a franchise." For a moment, I just looked at him and thought, "You must be as crazy as I am stupid."

How little did I thought of myself? As I think back on high school, I did become one of the most excelled students in my drafting class, as well as the first trumpet of the band. Yet, these accomplishments had absolutely no effect on my childhood belief about myself. The young lady I dated in high school, Jewell Johnson, was one of the most intelligent young girls in school; still, this successful friendship, along with all the other accomplishments, made no difference in how I felt about myself.

I did manage to break my father's direct influence over me by leaving Miami after working two months with Mr. Mack again. I was able to earn enough money to travel to New York and live once again with my mother. Mother and Dad had separated in Savannah back in 1949. Unfortunately, I did not break free of Daddy's psychological grip, even though I had made my getaway.

Upon arriving in New York, August 10, 1958, I got a job as a janitor at Roosevelt Hospital. After I had been working there about two months, they decided to train me as an orderly. Why not, this was in harmony with my belief of myself.

I became an outstanding orderly and found myself working on the night shift of the emergency ward. One night in 1960, we admitted a patient who had a heart attack. He was a successful forty-two-year-old black lawyer. As I undressed him, I recall my astonishment in seeing his underwear. I never knew people of affluence would wear such expensive-looking underwear.

That lawyer's heart stopped beating as we attended him, the intern on duty froze. I, without thinking, took over as I jumped up on the stretcher over the patient and started beating on his chest. I had seen other doctors do this in an emergency. I got his heart beating again as I yelled to the nurse to get ten cubic centimeters of levophed. I told the intern to inject directly through his chest into his heart. The nurse came back with the medication as the intern got a grip and composed himself to make the injection. We got that attorney's heart to stabilize before the resident doctor could come down to the ward.

That night after the emergency, I suggested to the nurse that we get a cardboard box and put everything we could think of that we would need for heart-attack patients. "We could put it all into the room where the heart flabellate is!" shouted the nurse. They all thought it was a marvelous idea, and we started putting our first heart attack units together so that we would no longer have to run around trying to look for things.

Days later, I visited that attorney patient on the private ward. We had a good talk and I just had to ask him, "How can I become successful like you?" The attorney took a long blank stare at me and said, "Don't worry, you will make your move in life." "But how can you say that? You don't even know me?" I replied. "How old are you?" he asked. "Twenty," I said. The attorney then replied, "I can see how intelligent you are; you just need time to develop." I thought to myself, "I have another one fooled." Somehow, I really wanted to believe that fancy-drawer guy.

After leaving the attorney's room, I, for the first time, started questioning what I thought of myself. This man was a lot smarter than Daddy. Could Daddy have been wrong about me? I recall the doctors and nurses of the emergency ward had me go take some test to get me into their hospital's nursing school on a full scholarship. I really did not want to take those tests, but they insisted. Well, I took those tests, and I did incredibly poor, due to my reading ability at the time.

I started wondering why people kept thinking that I was intelligent. Perhaps I just happen to have some kind of look that confuses them. One of the doctors really wanted to teach me to play chess. He was convinced that I would be an excellent player. Well, I always made sure I had things to do whenever things were slow on the ward. Especially, if I could see that he had time to teach me to play chess. You see, I had always heard that chess is a game for the intelligent, so I wanted no part of that. The universe always created a gracious escape for me.

As time went on, I started wondering just how my thinking compared to other men. Then I realized that the army would be a great place to make such a comparison. I have always heard that the army will make you or break you. Just how would I stuck up with other young men of the world? I wanted to know.

I went down to the draft board and signed up for the army. It took a few months for me to get my notice. As I waited, another orderly who worked at the hospital was trying to cut a record and got me interested in the record industry. I started trying to make my move like the lawyer said. I started writing songs and made a demo of one of my songs called "My Heart, My Body, My Soul." After pounding the pavement in New York's Tin Pan Alley for about six months, I finally got a music publisher interested in recording me. They were not so interested in my songs. They wanted to write the songs and record me as an artist.

It was then that I realized that I could become a star. But wait, I was not supposed to amount to anything. Yet, this company was calling and almost begging me to work with them. Cognitive dissonance set in, and I became frightened and uncomfortable. The universe, however, had already prepared another way out for me. As a matter of fact, the way out was already in motion even before I started pounding the pavement in Tin Pan Alley down on Broadway.

My draft notes finally came, and I was able to tell that publishing company that "I was drafted into the army, and there was nothing I could do about it." I still recall the disappointed look on those publisher's faces as I left their office. However, I felt so much better knowing that I would not become a star.

I requested to enter the army from Miami, Florida. I wanted to visit old friends and family before I went in. I flew down as I watched the winter weather change to spring. I spent a week with my father and family. During my visit, one of my cousins was driving me up to north Miami when, over in the right lane, I saw this old pale red 1949 convertible Oldsmobile with a familiar face in it. He was a bald-headed man with wire-framed glasses. As we passed, we looked at each other, and we started to smile. I shouted to my cousin, "Pull over, I know that guy." We pulled over as Mr. Mack pulled over behind us, and he and I got out of the cars. We were so excited to see each other. "Charlie, my boy!" shouted Mr. Mack, "Are you back to stay?"

I informed Mr. Mack that I had been drafted into the army and that I came down to see old friends and family before I go in. "Damn it! Charlie, if you had just stayed with me, you would have been a wealthy man now. I am taking Burger King all across America." I just looked at him with a puzzled smile, because I could not see how his success would have made me a wealthy man. We must have had quite different ideas of wealth. I knew even back then that I had to find my own way in life. Besides, being his boy was unacceptable to me. Well, we said our goodbyes and wished each other well.

Somehow, I feel that a person should make his own wealth through his contribution to humanity. As I think back to find where I got such an idea, I can only recall once when I was in the third grade when my regular school teacher, Mrs. Ford, was out sick for a week.

We had a substitute for that week, and on her last day as we were leaving class, the substitute teacher was standing at the doorway as we walked out. She pulled me to the side as everyone left. The young woman stooped down and put her arm around me as she looked me in the eyes. Then with a soft, yet serious voice, she said, "Charles, you are not working up to your potential. You know when you grow up, you are going to have to create your own thing in life, okay?" I responded by saying, "Yes, ma'am." The substitute then smiled as she kissed me on the forehead and then stood back up.

This substitute must have been a person with very high intelligence. She would have had to be to see above the grades I reflected. I sat in the third group, which was for the slowest kids in the classroom. Yet, she was able to recognize me as someone with potential worthy of her time.

Looking back, I guess she must have been able to see something special in my thought process or my artwork. I remember leaving that class thinking she has not been here long enough to know just how stupid I really am.

After getting into the army, the only job I tested out for was the infantry. Still, this was good for what I wanted to do, to evaluate myself with other men. After I had been trained as a professional killer in the infantry, I was sent to Germany. Now, I could start to make my comparison against other men. I needed to know who was right, Daddy or that attorney from the hospital.

Because I asked only this simple comparison from the universe, the universe gave me a perfect place to make that comparison. I wanted to see just how I staked up against other men of the world.

It took me about two months in Germany to realize that the infantry was not for me. I soon saw that I was not at all inferior to anyone in that infantry company. What's more, I no longer cared for those infantry people to be in charge of me. Without knowing what I was doing at that time, I just started praying as I asked ( ) to show me a way out of this mess.

As I lay on my bed after my prayer, the thought came to me to go to that band up the street and audition. The next day, I went to the band and got an audition. I told them that I had not played my trumpet since high school, but if I was accepted, I would be willing to work and bring myself up to par.

After the audition, it took a month and a half before my orders came for me to go to the band. During which time I spent my evenings in my room just daydreaming about things I thought I would enjoy having. There is a saying, "You should be careful of what you wish for, you may just get it." I used to fantasize about having a shiny, new black car; an attractive, blue-eyed, blond haired young German girlfriend; and becoming the first trumpet in that band up the street. I did not care if I could really have all of that or not; it was just a very pleasant way to pass my time. It was quite an enjoyable daydream. Little did I know that I was sending that out into the universe for its crystallization.

The day before I got my orders to go to the band, my infantry company got us all up at 3:00 a.m. and took the whole company on a long hike in the woods. The temperature was about twenty-five degrees, and it had snowed earlier. We all had on our cold-weather gear; we each had received a box of rations, and then they started walking us out of the front gate. We did not get breakfast. They just started walking us in the woods.

At first the walk was okay, but by 10:00 a.m., I was so mad at the whole company I just lost it. Our company commander and first sergeant were leading this walk of the A Company, Fourth Infantry Division, around a large snow-covered square opening for farming in the woods. The snow-covered clearing was about one thousand by one thousand square feet. I just started walking diagonally straight across the clearance to intersect the leaders of this insane walk in the woods.

With all the headgear we had on, the leaders did not see me until I was almost upon them. What got their attention was my squad leader chasing after me, yelling, "Get back in rank, Private!" "No," I replied, "I want to stop and eat, and I want to eat now. You cannot teach me to go hungry when I have food. If we did not have food, then that would be different. But we do have food, and I want to eat it now! I am hungry and tired." My squad leader said, "You are going to lose your stripe if you do not get back in rank."

By this time, the company had all stopped as everyone was observing this episode across the opening as we were closing in on the commander. I was so angry I started trying to pull that stripe off my sleeve. "Here, take it, take it!" I replied at the top of my voice. But the stripe had been sewed on too well. I was unable to rip it off. The company commander finally intervened by saying in a quiet, calm voice, "Okay, let us stop and all take a break and eat."

My squad leader and I went silently back in rank with our squad, and we all ate with a thirty-minute break. Then we started back to the company. No one in the company ever said anything to me about that incident. Well, no one other than my friend who was so surprised to see me take on such a character and do such a drastic thing like that. Even I was sure I would get some kind of punishment, but to my astonishment, I never got in trouble over that incident. Perhaps they were just testing the durability and breaking point of the men. Well, they found out from me.

The next day as I was out working on my tracked personnel carrier, my squad leader came to my vehicle and said, "The first sergeant wants to see you in the orderly room now." "Well," I thought, as my heart sank, "this is it." I thought, "Just go in there and take whatever they had to dish out like a man." To my surprise, the first sergeant said, "Go upstairs and pack up all your gear, you are going"—"Oh God, not to jail," I thought—"to the band," as he gave me my orders.

I found the men in the band a lot less aggressive. They spent a lot more time experiencing and enjoying the German culture. Their nails were manicured, and they just looked more affluent.

Many of them even had German friends. Many parts of the town, like Bamberg, were off-limits to the GIs, but not off-limits to the band members.

One of the bandsmen asked me, "Now that you are in the band, what do you hope to accomplish here in Germany?" I thought to myself, "Compare myself to you guys, for one thing." However, in trying to give an acceptable answer, I thought further and said, "First, I am going to become the first trumpet of this band, as I am learning German.

"Then I am going to get a new car, so that I can travel away from this city. I want to find myself a beautiful, young, virgin, blue-eyed, blond German girlfriend. One who could not speak any English, one who was not spoiled by the GIs. What's more, she is going to love me truly and unconditionally."

The band members just looked at me, then at each other, then they broke out laughing so hard; some were on the floor. Then they started explaining to me that Flu Ellen and Harvey were the two first-trumpet players, and they were not going anywhere for a year. They further explained that in order to get a car over here, I would have to be at the grade of sergeant or better, and that there was no way I could make sergeant in this band. They also made it very clear that as for the blue-eyed, blond, virgin girlfriend, forget it. Only the prostitutes down on the Strass were available to us.

Well, at first, I believed them. I went back and lay on my bunk feeling a little abashed. Keep in mind that this was 1961. I allowed my good buddy Boone to take me down to that segregated part of town one evening. Boone set himself and me up with two of those prostitutes; I was so repulsed after entering into a verbal contract with her; I was unable to perform.

I recall her saying, "You mean you can't get it up with all this beautiful white meat?" That did it. I demanded my money back. That was the first and last time I went down on that Strass. That was also when I decided that I could not, and will not, deal with another prostitute. I decided never to go down there again. Segregation in the towns still existed in Germany, even though the army had been integrated.

I started staying in my room at night as I went back to my very enjoyable daydreams of what I really would like to be doing during my stay in Germany. Because my daydreams were true, honest desires and wishes, I was, unknowingly, sending out waves of quantum energy into the universe, which was creating that reality for me. Certainly, knowingly or unknowingly, I was changing my universe. The quantum energy I sent out had immediately started crystallizing into particles of possibilities.

Out of the six trumpets in the band, I sat sixth chair, third trumpet. Now I knew from high school that in spite of my stupidity, I was first-trumpet player. I started getting up two hours before everyone in the band got up each morning and went up into the practice room. Within six months, I had become the first-chair trumpeter. Systematically I had successfully challenged each trumpet player for their chair, well, with the exception of one.

There was Bernard Jackson (Bennie) from Norfolk State University who played second trumpet; when I moved up to play second trumpet with him, I said to him after one week, "Bennie, I think I am ready to challenge you." Bennie, having sat next to me, had become intimidated as he could clearly hear the preciseness of my articulation and the brilliancy of my tone. I, on the other hand, was able to hear the many mistakes he was making as we played. Bennie did not want to challenge me. He shouted, "You want my chair, then just take it!"

With Bennie now sitting to my left, I started preparing myself for the challenge of Harvey who played first, but who was not the soloist. Harvey was, by far, the most challenging one of all. You see; Harvey knew that if he lost his chair he would be playing second, and no first-trumpet player wants that if he can help it.

I went into town and bought myself a brand new Selma trumpet with nice responsive valves. I also picked up a copy of Harry James' "The Flight of the Bumble Bee." I learned to play that thing at a tempo of 180. One morning as the band came up for rehearsal, I started playing it. This really got Harvey's attention. But Harvey refused a challenge with me. He was able to do that, because he did outrank me.

After four months, however, I went to the band director and demanded the challenge. Harvey was angry, but the challenge was set for a week later. For the first time, I saw Harvey really practicing his trumpet. Of course, I knew this could only come down to a sight-reading challenge, so I got out my Arban book and tried to sight-read everything in the back of that book.

A week later, at eight in the morning, the band assembled in their seats, and Harvey had me start out with "The Barnum and Bailey's March," a favorite of his. The band gave it to me. Harvey and I challenged for one and a half hour before he finally gave in to me.

Men in the band started showing respect to me, and for the first time in my life, I started feeling as if I deserved my accomplishments. This band, however, was what was known as an unauthorized band. This type of band was authorized only through the authority of the post commander. Good musicians that they drafted into the army were sometimes given clerk's job, etc. The post commander, however, had the authority to reassign any qualified musician to his post band.

This was an excellent place for misfits like me. This band was such a contrast from the infantry. I could look down from the third-floor window of my room and see the men in the infantry company I just left, working on their armored personal carriers. I was one of those drivers who had to be out there every day. This too was one of those cold German December days. I could hear and see the sergeant yelling at them.

As a member of the band, we would get up at seven and go to breakfast. We would then start band rehearsal at eight each weekday. Rehearsals usually lasted until nine thirty, at which time we would take a break for coffee until ten. Then, back to the rehearsal until eleven thirty when we would break for lunch. At one in the afternoon, provided we did not have a job to play for that day, we would be free to do whatever we wished.

I signed up for an afternoon German class at the education center and started going to the library to try answering some of the SAQ which had always troubled me. Even as slow as I read, I had time to get through some books on atoms and the universe.

With so much time on hand, I got a job at night at the NCO club as a waiter. In a few months, I had saved up enough money, from just tips alone, for the down payment on a new car. What's more, the servers enjoyed teaching me many wonderful, polite German phrases. I was learning German very fast.

I was now able to afford my new car. But as a private, there were still obstacles to overcome. Believe it or not, this was one of the easiest of all to overcome. I put together a small rhythm-and-blues band, which I used to justify my special request for a car. I just made a request to the post commander asking for permission to have a car, so that I could provide transport to haul our musical instruments around town in order to provide entertainment for the black soldiers who had to patronize the segregated part of Bamberg.

The post commander granted me permission, and I got my new car, even though I was only a private first class. Now I was able to drive away from that GI town after work and meet wonderful German people who did not have such a bad attitude toward the GI.

Remember the saying, "Be careful of what you ask for, you may just get it"? Well now, because of my childlike mind, I had allowed the media to form my sense of what was beautiful. At that time, they were implicating and downright promoting the blue-eyed blonde. Well, of course, I thought I wanted one too.

I had a car and somewhat of a command of the German language, so it was not long before I became friends with a German guy named Otto. Otto just happened to have a young sister-in-law that fit my dream girl's description. You must understand that I never even mentioned to Otto that I wanted to meet a girl. Yet, not more than two hours after Otto and I had enjoyed a couple of beers, he said, "I know someone that would be the perfect girl for you."

"So, when do you think I could meet this perfect girl?" I asked. Otto drank down the rest of his beer and said, "Right now." "Right now? Shouldn't we call and make an appointment, or tell her that we want to come over?" With excitement, he replied, "*Nein, nein, auf, geht zu! Wir gehen!*"

Otto directed me through the small streets of old Bamberg to his in-laws' house. It was a three-story apartment building. We rang the doorbell downstairs and entered a hallway with a staircase that led up the three flights. Otto and I were able to look up to the third floor where we saw Ann looking down to see who was coming up. I was able to get a quick glance at her bushy french-twist hair that she had taken apart for the evening. She yelled as she ran back into her door, but, by that time, Otto and I had run to the top of the stairs and knocked on the door. Ann had made her transformation. She opened the door, and to my surprise, she had put her hair back up into her french twist and had put on some lipstick as well. I thought it was perhaps another sister. As we met, I could see that she was elated to meet me, even though she spoke no English. What's more, she had all the other attributes I wished for.

Ann's father was a little old man with soft, gentle blue eyes who was old enough to be Ann's grandfather. Her father took to me at once and started treating me like a proud dad would. Ann's mother, on the other hand, took a "wait and see" attitude. They invited me back for dinner that Sunday, and Ann and I started dating just like that.

Ann, with her unconditional love, became my companion for the next twenty-three years, because I did not ask for anything in the area of personality. The universe did not consider our personalities clashing when setting up the quantum mix. It would have been so easy to have also dreamed that she would be my very best friend. So you can trust me as I tell you to consider everything when you dream your dreams. Be careful as you visualize it all.

In summary, to manipulate quantum energy, consider carefully what you want, enjoy the dream as if you already have it. Enjoying the dream is like having it; desire and enjoyment are the driving forces behind the wave that becomes the particles of reality. It is just that simple.

# CHAPTER 7

## Developing Confidence

By now, some people are wondering how a person could, with such a devastating childhood, be able to overcome and develop the confidence to attempt the answers to such questions. I would have to say that it all started when I went into the army in 1961. I met my most memorable army buddy, Bedford Boone, at Fort Bennings, Georgia. Here was a little guy who looked a lot like me in many ways. The noticeable difference about us was that I was soft-spoken and somewhat meek, while Boone displayed all the confidence in the world. He could be loud, overbearing, and aggressive. It was like seeing me in a different mental form.

Boone and I became the best of friends; we were like twin brothers. He had an aggressive quality, which I admired; I had an intelligent, gentle quality, which he admired. His mother, whom he described as a harlot, brought him up in the slums of Cleveland, Ohio. He had no respect for women and very little for men. Boone was what you could call a street survivor. Another major difference in us was that where I would think first before acting, Boone would strike first and think later.

Once, when we first started hanging out together, some drunken guy came up to me insisting that I give him money. I told the guy that I did not have it. The guy aggressively got up in my face with his demand. Boone saw what was happening and came rushing over, pushing the drunk back so hard; the guy fell to the floor. Boone said, "Look, you mf#!*er, he said he don't have it!" That guy got up from the floor with fear in his eyes, apologizing as he backed off. "See, man, you have to learn to take mf#!*er's confidence like that if ya gonna survive in this world!"

I used to think the main reason we liked hanging out together so much was that we both loved girls. What's more, we were living in a time when the only thing there was to worry about was not to impregnate any one . . . well, that, and keeping Boone from getting into fights with other people.

Basic training proved quite a challenge to me. I remember wanting to just give up. The basic training leadership was a master in breaking one's self-worth down to nothing. As you know, I never had much self-confidence in the first place, and the army chipped away the little I did have.

After our basic training at Fort Benning, Georgia, we went to the advanced infantry training at Fort Hood, Texas. Boone and I were put into different platoons, and I became acting squad leader after the first month there. Boone, in another platoon, was always getting in trouble with everyone.

One Sunday evening a week before we were to graduate, Boone and I were at the service club enjoying the company of some young women who had come on post with the USO. We were having a great time with the girls, when I reminded Boone that we had to be back at the company by six that evening for a GI party (a gathering to clean our living quarters). Boone said, "Look, man, what's going to happen if we are not there?" I replied, "For me, it will be the end of my leadership as squad leader." "So what, you will be out of here in a week man!" Boone shouted. "What do you want to be a squad leader for anyway? Come on, man, screw it, let's stay here and enjoy ourselves." I thought for a moment and said, "Well, okay, but what about you? You could be in deep trouble." "Man, Coop, do I look like I give a shit?" I looked at Boone as we both started smiling. "No," I replied. Boone pushed me back slightly on my right shoulder as he said, "Okay then, let's stay and have some more fun."

The next morning, I was no longer squad leader. Boone, however, got restricted to the company area until we graduated. Having to move back down to just a squad member did somehow put me back into a more comfortable zone. To tell the truth, I never did understand why the platoon sergeant had made me squad leader in the first place. However, I did come to one conclusion; I realized that I should never follow Boone's judgment again.

Upon our graduation, Boone and I were sent to Germany together on the same orders. After taking a full forty-five-day leave, we went over on what was known as a troop-carrying ship. This was an eight-day trip; this was not what you may call a fun cruise, but we made the best of it by creating our own fun. Boone thought that we should put some of that low crawl training we learned into practice by crawling into the ship's bakery one night and stealing some sweet rolls. This was quite an experience for me, because I had never taken anything in my life.

Well, we had a great laugh as we ate those rolls, but I told Boone that I did not want to do that again. Well, Boone kept doing that each night on his own until he got caught and was thrown into the ship's brig by the commanding officer. The brig (jail) was in the very bow of the ship. They kept him there for three days. He really looked bad when he came out. It was then that Boone told me that he was always in trouble before he came into the army. He said that the judge gave him the choice of coming into the army or going to jail for all the trouble he was always in.

I made it clear to Boone that day that if he and I were to continue our friendship, he would have to agree that he would listen to me for his own good. We shook on it, and that became our mode of operation.

Many people in the army often wondered and sometimes even asked us, "What do you two see in each other?" I cannot recall a time when Boone and I thought to discuss the matter. We were truly an odd pair. Now that I am looking back on my past, I can admit that I liked his confidence. I enjoyed watching him take other people's confidence away with his aggression. Boone referred to me as Mr. Ambassador and to himself as the enforcer. We just had the greatest admiration for each other. Boone would give me the shirt off his back along with his very last nickel if I needed it. We could trust each other with our lives.

We arrived in Germany in the September of 1961, but by December of that year, I had left infantry for the band. I was no longer with my twin and good buddy. It was not long before I started missing his display of confidence. I guess I had somehow started living off his confidence. After about eight months apart, Boone came to me one evening, almost in tears, telling me that he would be going to jail if I did not help him to get out of the infantry. He had already lost the one stripe he made before I left infantry.

Boone did not play a musical instrument, but the cymbals player was rotating back to the States, and I was sure I could teach Boone when to clash those silly things. But there was still another big problem with Boone coming to the band: no one in the band liked Boone. They had seen him in action downtown, and they did not want him around. I explained that he never acted that way when he was with me.

Everyone in the band was quite adamant about keeping Boone out. Still, I could not see Boone stay in the infantry and go to jail. So for the first time in my life, I pulled a power play. I told the band director that if they did not let my partner in the band, I would go back to my infantry company to help keep him out of trouble. You see, I truly cared about what could happen to Boone.

By then, as the first trumpet in the band, I was well established. I knew that they could not do without my leading tone. After all, we played mostly military marches, where the intro was usually strongly relying on the first trumpet. The director's downbeat meant nothing if I did not come in.

I took advantage of my position for the first time in my life because of my desperation for getting Boone out of infantry. I told the band director that if they did not let Boone in, I would leave the band and go back to my company to keep my buddy out of trouble.

The band had an informal meeting without me, and they decided that they would prefer letting Boone in rather than letting me go. They knew they had no choice, even though they hated Boone. The director came back to me and made me promise that I would be totally responsible for Boone's action. I knew that I was the one person in the whole world that Boone respected, and he would follow my wishes.

I went over to Boone's company that night and told him to lay real low and that he would be getting orders to come to the band very soon. He was elated, even though he had to accept certain conditions with coming to the band. "Boone," I said, "I want to tell you the truth. No one other than me wants you in the band. They are only letting you in because I threaten to leave the band if they did not let you in."

"Man, you did that for me Coop?"

"Why, of course, good buddy," I replied. I watched Boone's eyes glassed up with some rare tears as he threw his arms around me in hopes of masking his emotion.

After a minute, we pushed each other away, smiling, because we were not supposed to be acting like that. "Man, Coop," shouted Boone, "No one in this world would have done this for me. Thank you, man!" "Boon, if you really want to thank me, you must promise me that you will listen to me when you come to the band. The band director told me that I would be held personally responsible for your actions."

*A German beer festival with Boone and I*

Boone came to the band within four weeks, and we became twins again. I felt a sense of power through my abilities and with Boone on my side. Not long after Boone came to the band, our bandleader rotated back to the States. The person who took over the band was not a musician. He was one of those dogmatic (E-7) sergeant first-class types. He obviously went over to the base commander's office and convinced the colonel that he could discipline the band as well as direct the music. As you may recall, my daddy was one of those dogmatic, disciplinarian types; this was a formula for trouble.

His name was Sergeant Polite, but there were nothing polite about this man at all. During our rehearsals, he learned to follow the band with his baton by waving it up and down. He had to learn when the march was ending. The march would end, and he would still be waving his arm. He could not even read music. During this time, I guess he was sizing up the band members and deciding whom he would respect and whom to push around.

Somehow, he saw Boone as a man to be respected and me as someone with whom he could push around. I guess he needed to have someone to pick on to show off his idea of his manhood. He could see that I was not the John Wayne type. So, he started picking on me every chance he got. What's more, he was becoming downright rude to me.

After Sergeant Polite had been with the band for about thirty days, he was then faced with his first major band commitment. We had to play for a division football game. Bamberg's post football team was playing Wurzberg's team. The colonels from each division along with their entourage were to be there in support of their teams.

As the band fell out for the march to the stadium, Sergeant Polite made his inspection. When he got to me, he said, "Cooper, get your sorry ass back upstairs and get that fingerprint off that head gear. Move your ass now!" I took my hat off, rubbed it on my trousers, and put it back on. "Cooper, I said get your ass upstairs and clean it." I was so humiliated as I ran upstairs and then ran back down just to please him. I guess he wanted to show the band that he was the absolute ruler.

I resumed my place in formation, and we marched to the stadium on drumbeat. It was just as well because I was too angry to play. We sat up in the stands about twenty-five feet from the colonels and all of the hi-ups. Sergeant Polite knew that this was his day to shine and show the colonel what a great job he was doing with the band. Now it was time to play our first march, "The Colonel Boggy March." Sergeant Polite raised his baton and gave the downbeat before I could get my horn to my lips. The band went *squeak, squeak,* and died out within two measures.

I then realized that that dummy never looked at me before he gave the downbeat. I still felt humiliated, so much so that I did not care if he made a fool of himself in front of everyone. Sergeant Polite raised the baton a second time and gave the downbeat. This time I deliberately did not play. I then saw a chance to humiliate him as well.

Again, the band could do nothing more than *squeak, squeak, squeak,* and then nothing. Sergeant Polite yelled out, "Come on, guys, what's wrong with you?" I heard one of the band members who was in front of him saying, "Cooper did not come in." Sergeant

Polite yelled, "Cooper? What the hell does Cooper have to do with anything?" Then I heard another band member who was closer to him whispered, "He's the first trumpet." Sergeant Polite looked dumbfounded for a second as he looked over at me, then in his *marché* voice said, "Let's go, Cooper," as he brought the baton down. I just looked straight at him with contempt as the band went *squeak, squeak,* and died out for the third time.

Sergeant Polite looked at the colonels as they were all staring at him. Then he looked over at me, and I could see fear in his glary eyes. Now I just stared him down with the utmost contempt and anger in my eyes. I wanted him to feel what it was like to be humiliated. As we stared at each other, I began to really see a glassy-eyed, fearful person staring back at me as he said in a meek, soft voice, "Cooper, please."

It was then I realized that I had taken his confidence away. Then I felt a little sorry for him. If you are unfamiliar with the military, let me tell you this: you don't embarrass the base commander and expect to walk away free. Sergeant Polite was about to get the blame for embarrassing the colonel in front of his guests.

Sergeant Polite gave the downbeat once again, for the fourth time, and with my leading tone, I gave him a strong, loud introduction. You could see the relief on his face after I started playing. When we got back to our barracks, he asked me if I would come down to his room for a talk. When I got there, he displayed a politeness like I had never seen before.

Sergeant Polite said, "Come in, young man," as he put his arms around my shoulders and escorted me to a seat. "Have a seat." He then went over to his little drink cooler and opened it, displaying all of his soft drinks. "What would you like, young man?" "I do not want anything from you," I replied.

"Cooper, I asked you down here because I want to know why you did not play out there." "Sergeant Polite," I replied, "When you gave the downbeat the first time, I could not get my horn to my lips in time. But the second and third time, I wanted to show you that you must look at me to be sure that I am ready before you give the downbeat."

Sergeant Polite said, "So the band needs the first trumpet to come in?" I just looked at him and thought, "Does he think the clarinets and flutes are bringing in the band? Man, this guy is even more ignorant than I thought." I replied, "That's correct, Sergeant. Most marches rely heavily on the trumpets for the introductions, so if you do not want to be embarrassed, you must look at me before you give the downbeat."

Our conversation ended with smiles and handshakes. I left his office wondering if there was only a certain amount of confidence in the universe. If I wanted some, I would just have to take it from those who had it. From that day forward, Sergeant Polite made sure that I was happy in the band, but most of all, he made sure I had no ill feeling toward him. I felt that I had taken enough of Sergeant Polite's confidence to meet my needs.

In late 1962, the Seventh Army commander, General Clark, started disbanding the division's football teams, the Seventh Army Symphonic Band, and chorus, along with many of the other nonessential military expenses. Even the Fourth Infantry Division post band was being disbanded, and we were given a chance of audition for the authorized division band or for the Garry Owens Dragoon's unauthorized band down in Nuremberg.

I chose the unauthorized band down in Nuremberg, Germany, because one of our ex-trombone players, who had previously left our band to go up to that Wurzberg division band, informed me that, that was a bad choice. His name was Cranes, who told me that he hated it there; they had to pull KP (kitchen police) as well as guard duty. He also said that the sergeants were tough.

Well, of course, I asked for Nuremberg. Within six weeks, I got my orders to report to Nuremberg band that same night not later than 24:00 hours. I am sure my infantry company commander had those orders in his office for at least thirty days. I only had hours to clear the post, pack, and drive the fifty miles down to Nuremberg.

Boone, on the other hand, was not able to pass the audition; well, it did not matter much anymore, because his tour of duty was nearly up. He was only in the army for a two-year enlistment. I, on the other hand, still had eighteen months to spend in Germany.

The director of the Nuremberg band was a young black alto-saxophone player whose name was Jessie Taylor. Jessie had a master's degree in music from Tennessee State University. He had teacher's credential as well. Jessie was qualified to teach music at the college level, yet he only held a low-pay grade of an E-4. With his education level, he should have been at least a major. That's the way it was for draftees at that time.

Jessie would spend his weekends playing in a jazz keller in downtown Nuremberg with his German musician friends at night. He would also play in his room during the day along with recordings from John Coltrane, Miles Davis, and Canon Ball Adlai. Such was the caliber of his alto-saxophone horn playing. Jessie was one of those few people who had perfect pitch.

Because I was the only one in the band that had a car, Jessie latched on to me. Within four months, however, Jessie's time was up. He moved to Sweden and married a Swedish girl whose father was on the board of education in Stockholm. Jessie got a job teaching music at one of the universities up there.

After Jessie left, the band's leadership was up for grabs, but none of us wanted some noncommissioned sergeant coming in and trying to lead us. I certainly did not want that; I had already experienced what that could be like. One of the trumpet players, who came into the band with me from another disbanded band, thought that he could run the band. Specialist 4 Jenkins, a bright twenty-year-old fellow, went over to the colonel's office and made him aware that he had three years of college and a GT (general aptitude) score of 145. This was enough to convince the colonel to make him an acting sergeant with the power to govern the band as he saw fit. Of course, Jenkins was one of us, and we told him that we want the band to be run democratically or we were not going to cooperate. Jenkins had no choice but to accept the input from the other three leading musicians.

We called ourselves the four nice guys. We were the executive board for the band. Jenkins was Irish; Peter Gasparetti was Italian; Freddie Goldstein was Jewish, and I represented the black interest. Without the four board members' approval, nothing was to take place.

Here was what we, the board, scheduled and agreed upon for the band. First, we no longer wanted to stand in the morning formations just for accountability. We decided that we could take our roll-call when we come to rehearsal each day and turn it in to the company by 8 a.m. In reality, we just did not want to walk around and pick up cigarette butts and papers off the ground, as the company did after each morning formation, especially since none of us smoked.

Rehearsal would start at eight each morning. If someone were late, they would have to answer to our executive board and not the company commander. From 9:30 until 10 a.m., we would break for coffee, and then from 10 a.m. until 11:30, we rehearsed once more. We took lunch from 11:30 until 1:00 p.m. After lunch, if we had no band commitment for that day, we would get our assigned military bus and take the band out to some civilian activity in Nuremberg.

Because I was the only one in the band with license to drive in Germany, I drove the band's bus. I spoke southern German and was able to get along quite well with the locals. You could say we had it made in the band.

On 16 November 1963, I married Ann, the girl I had created in my mind from Bamberg. Oh, if only I had known at that time, I could have created a best friend as well.

After being engaged to Ann for a few months, I could see that our personalities were not right for each other, and I did not see where we had anything in common either. Despite this discrepancy, she refused to let go. What's more, at that time in my life, I just did not have the confidence to refuse someone who knew what they wanted; she knew with clarity and desire just what she wanted: marriage.

When Ann and I rotated back to the States for discharge in July of 1964, I saw no other choice but to reenlist for the bonus by taking six more years. Why not? The band had been great and I needed the money to set up household, because our baby was on its way in August.

At that time, I still had not realized that all I had to do was ask the universe for whatever I wanted and I could have it. I never asked for much from ( ). Nevertheless, because Ann was only seventeen and spoke no English, she was totally dependent on me; I had to start asking a few things of the universe; to never be separated from my family during my twenty years in the military was one thing I asked for.

Not to be separated from one's family was unheard of in the army and was considered unrealistic. Yet, that was all I asked for, other than to make at least the pay grade of E-7, so that I could support my family. Would you believe it? For the exception of a four-month tour I volunteered to spend in Babcock, Thailand, on a top-secret mission, we were never separated. Well, not even then, because she came over with our two kids after two months, and we lived like ambassadors for the remaining two months.

Back in 1964, after I took those first six years, we were sent to Fort Lewis, Washington, where I was assigned to an authorized division band for the first time. I did not like the structure at all. Fortunately, they made me the post bugler; all I had to do was drive down to the flagpole at 6:00 a.m., with my wife and kid in the car, and play bugle calls. Revelry was the first bugle call to start the day, and the playing of taps at 11:00 p.m. was the last bugle call of the day. In all, I was at the flagpole four times throughout the day. It was all right; I had no other duties other than to play all the calls on time.

In 1965, the complete fourth division was put on alert to go to Vietnam. No one was allowed to transfer out of that division. I got orders to go to the school of music in Norfolk, Virginia. Imagine that, the whole division along with the band was going to war, and I was the only one being sent in the opposite direction!

So why was I being sent in the opposite direction of the war? What higher power was working in my life? The answer is none other than the power of the quantum connection, a connection that we are all a part of. If you do not clearly ask the quantum connection for what you want, you will wind up with some of what you do not want.

When we got to Norfolk, I learned that it was illegal in Virginia, at that time, for whites and blacks to marry. Now I found myself in a dilemma, and I could see no other choice but to just go up to the Pentagon and ask to be stationed somewhere where my family and I could live legally.

"Just where would you like that to be?" asked the major who was in charge of all band personnel armywide. Back to Europe would be a place I know we would not be breaking any laws for our marriage.

The major explained to me that I could not be sent back to Europe until I had been stateside for eighteen months or more. The major then tried to convince me to leave my family in Washington, D.C., and commute to and from Norfolk every weekend. I made the major aware of Ann's age and that she spoke no English and could not conduct any business without me at this time.

"I can see no way to help you other than to let you out of the army," replied the major. I shifted my focus to look out from the office into the Pentagon hall to see Ann pacing to and fro, pushing our cute, little, well-mannered eight-month-old son in his stroller. Each time Ann would pass the door, my little son would smile at me.

I turned back to the major and said, "My obligation to my family could not allow me to just leave them on their own." Sir, I am going to take my chance on the outside. But what will you do to support them? I do not know; I just know that selling shoes would be better than leaving them to fend for themselves here in D.C.

There were two civil servant women sitting at their desk overhearing the whole conversation, and when the major said that he was not able to do anything for me, those two ladies, one black and one white, looked at him with daggers in their eyes. If looks could kill, that major would have been lifeless.

To my surprise, the major quickly walk over to a tall file cabinet saying, "But wait, maybe there is something I could do. I do know of a band that I may be able to send you to in New Jersey." The two ladies, never saying a word, bowed their head back to the work on their desk, as the major pulled a band file and gave me a verbal order to report to Fort Mammoth. He told me that my orders would be following soon.

We left the Pentagon ecstatic, knowing we would not have to suffer a hardship. The next day, we reported to Fort Mammoth, made our presence known, took a hotel, and waited for orders to sign in. The band had enough trumpets, so the director put me on French horn, which I played for only six months.

I got orders after only six months at Fort Mammoth to go back to Europe, to Vicenza, Italy, which was only thirty miles south of Venice. Italy turned out to be very good for me; I even thought, at one time, that if I were to retire in any other country other than the States, it would be Italy. Unlike Germany, the Italian people really made me feel equal to them. I never felt that the Italians looked down on me. We lived on the Italian economy, and I learned to speak adequate Italian during my three years over there. I really started feeling good about myself there.

The rebirth of my life, shared in chapter 3, was in Italy 1966. This was also where I first started making notes and sketches of thoughts as I imagined them to be. I decided that when I rotated back to the States, I would change my military occupation skill (MOS) to an illustrator.

When we got back to the States in 1969, I still had six months left on my enlistment. We were then sent to Fort Ord, California, to finish my last six months. During that time, I took some of my drawings over to one of the graphic shops on the post and was accepted as a draftsman.

June of 1970, my enlistment was up, and I had made up my mind. I got out of the army and enrolled in college. We moved to Palo Alto, California, where I entered Foot Hill Junior College. The cost of living while trying to go to school, even with the GI Bill, proved too great for us. By December of that same year, I had to join the army once more.

I was assigned to the Fifty-second Army band down at Fort Hood, Texas. We still had some of our savings left over from Italy and were able to buy a three-bedroom rancher. It was a new neighborhood with no trees other than the one small required tree in each yard. I wanted to line the whole block with shade trees, so I went to everyone on the block and surveyed for interest. Everyone thought that shade trees would be a good idea. I tested the soil and did my research for the best tree and cost. Everyone paid up front for their order, but I later learned that because of the volume of trees I would be purchasing, I was able to get them wholesale if I was willing to rent a U-Hall and drive to San Antonia to pick them up.

After talking with a few of the men in the band, I found out that they were eager to make a little money and help me plant my trees. Getting the trees wholesale gave me enough money to cover all my expenses and some profit to pick up some extra trees for the next street. Voila, just like that I had started a business. I named the company Budget Landscaping Inc. Before I knew it, my backyard was full of wholesale trees and plants.

I was now becoming a young man of confidence. I started designing lay outs of landscapes for apartment complexes as well as for homes. During my business activities, I came across a nursery owner who wanted to retire. The owner asked me if I wanted to buy him out and take over his business.

Was this the move that fancy-drawer lawyer said I would make in life? I had never thought about a business in landscaping, and I certainly did not ask for it. But here was this thing, with me knowing very little about plants.

I knew that the location just outside of the Fort Hood gate was prime, and the price was great for such a piece of property, but where was the money going to come from? Then a few days later, the first Burger King came to Killeen, Texas. I went to the franchise owner of that restaurant as it was being constructed in hopes of getting another landscaping contract. I introduced myself and asked the owner, "Did James McLamore still control this corporation?"

The owner said, "How do you know Jim?" "I used to ring a dinner bell for him when I was thirteen years old," I answered. The owner was so shocked he had to sit down on one of the sawhorse and just stare at me as he smiled.

"So you were that little guy ringing that bell!" the owner went on to say, "Yes, I remember you out there. That was before Burger King existed." I then asked him, "Do you know Mr. Mack?" He replied, "Oh yes, I used to play golf with him. But he has gotten too big for me now. Wow! You should give him a call. I bet he would be excited to hear from you."

The franchise owner stood up from the sawhorse, reached for his wallet, pulled out a business card from the corporation, and excitedly gave me the phone number to the Burger King headquarters. I thanked him and said that I will call him as soon as I get back to my office. Of course, I also did the landscaping of his restaurant.

I called Mr. Mack at the corporation later that day, but I was unable to speak to him. They asked me to leave my name, number, and a message. I told the secretary to tell Mr. Mack that "Dinner Bell Charlie" called. I left my home or office number as well as my army-band work number.

After about three days of not hearing from him, I began thinking that maybe Mr. Mack felt that he had gotten too big to talk to me too when, on the fourth day, Mr. Mack called me back during one of my practices.

The first sergeant of the band came into the practice room with a puzzled look on his face, as he said to me, "Cooper, the CEO of Burger King Corporation is on the phone and wants to speak with you." I jumped up with excitement, put down my horn, and headed to the office. The first sergeant followed close behind and closed the door after we entered the office.

I picked up the phone and said, "Mr. Mack?" "Charlie! Son-a-ma-gun! How the hell are you?" he asked. "I am fine, and how are you, Mr. Mack?" "Charlie, I am doing just great." I then asked, "Do you still have all those Burger Chefs?" Mr. Mack replied in a not-so-happy voice, "Burger King, Charlie, it's Burger King." I then, in a somewhat apologetic voice, said, "I know, but it's just that we have so many Burger Chefs in this area, it's the first thing that comes to mind; they are all you see around here in Texas." There was a pause.

"I understand. So, what have you been up to, Charlie, my boy?" "Well, as you know, I am still in the army, married with two kids, and I have started a landscaping business on the side. I know you must be busy and rolling down there, but I would love to come down and see your operation."

"No, Charlie, not right now, things are in a real mess around here at this time." Mr. Mack never explained if the mess was a renovation mess or an acquisition mess. As I wondered shortly about what kind of mess, another pause occurred.

"So tell me, Charlie, what can I do for you?" "Well, there is a nursery in this area for sale, and I was wondering if you could back me in getting it?" "Charlie, of all the black people I have helped, I would rather help you most of all."

There was still another complete silence as I thought to myself, "Why did he have to say it that way?" Then I realized that he still see me as that little stupid black boy. Now that really pissed me off. I started feeling unhappy and belittled by his statement. I started thinking to myself that that was most likely why he did not want me to come down to visit him. He is ashamed to have me at his office.

Then Mr. Mack broke the eerie silence by saying, "Tell you what, Charlie, write me a letter laying out your finances and what you really want to do." "Well, thank you, Mr. Mack!" Then Mr. Mack said, "Better yet, Charlie, write the letter to my house. Got a pencil?" Mr. Mack then very carefully gave me his Coral Gable home address.

I did not write him back right away. I was no longer happy with him. Even though Mr. Mack wanted to help me get some footing in life, I began having real problems with him, the longer I spent thinking about him not wanting to be seen with me. He still thought of me as that little dumb guy with whom he had to utilize to get where he wanted to be in life. The more I thought about him, the angrier I became. After about five days of thinking, I made up my mind that I just wanted to tell Jim just how I felt.

Sure, I could have written him and led him on by just giving him the information he asked for, but it became more important to me that I let him see that I was an evolving human being. If he had trouble with me in the way that I was growing and changing, then I did not want his help.

I wrote him a ten-page letter telling him all about me and what I thought. Furthermore, I decided to disobey him by sending the letter to his corporation. How dare he be embarrassed of me.

Telling Mr. Mack what I thought was something I was never able to do when I was a kid. I recalled in 1954, when they were starting to integrate the schools, everyone was talking about it, even Mr. Mack. One day, I went into the office to get my dinner bell where he was sitting at his desk reading the *Miami Herald* newspaper. He turned to me and asked, "What do you think about them integrating the schools, Charlie?" I said, without much thought, "I think it would be nice if everyone could get an equal education together."

Mr. Mack became outraged as he shouted, "But you people are not ready!" I looked at him and saw how angry he was and thought to myself, "Well, how can we ever get ready if we can't get an equal education?" Instead of answering him, I got my dinner bell and my chef hat and got out of the office. That was the last time Mr. Mack ever asked me what I thought about anything else.

Before I wrote to Mr. Mack, I started recalling the racial climate in the south in 1954 and how saying the wrong thing could upset him. I had to try to understand why he became so angry with me about my view on integration. Then it hit me. He had two little daughters who were about three and five at that time. Mr. Mack was a tall, gentle, golf-playing, well-dressed, Presbyterian, Irish man. Maybe the thought of his daughters in class with someone like me was just too much for him.

The more I thought about him wanting to help me because I was black and not because I was a key employee from his past, the more upset I became; I never saw him as a white man. To me, he was the man with the plan, and it really hurt that he saw me as just a black kid. It hurt to the point where I no longer cared if he backed me or not. It became more important that he heard what I thought.

I remember writing him on a yellow legal pad. "Dear Mr. Mack, I am writing this letter to you in pencil because I want it to flow straight from my heart. Did you notice anything unusual about me when I was out there ringing and flipping that dinner bell in rhythm for you?

"What you didn't know was that those people who would stop and talk to me were songwriters, inventors, and business people who loved to talk to me and get ideas. I suggested plastic grass to one, a motorized surfboard to another, and many song ideas to songwriters, yet you saw me as quite inferior to you. Well, to tell you the truth, so did I."

In the letter, I went on to describe my episode at the education center in Italy and the impact it had on me. I even drew a diagram to illustrate my thinking about human development and our hierarchy of needs. I wanted to show that after we (human beings) have satisfied our hierarchy of needs, we could continue the expansion onto an even higher level if desired. We could attract our wishes through desire and even communicate through our thought waves.

After about seven pages into the letter, I began realizing through my writing that what I really wanted out of life was to somehow do mind research. I told Mr. Mack how fascinated I was about thoughts and about how I thought they worked in our minds. I also told him that when I get financial freedom, I was going to spend my time doing mind research.

Well, I never got the backing for the nursery. Mr. Mack took a long time before he answered my letter. During the wait, Ann made me feel quite bad for writing and telling him my thoughts. She said that Mr. Mack did not want to help a crazy-thinking man. Still I felt that it was more important that he changed his image of me if I was to accept his backing. I decided that I did not want his backing if he could not accept development.

I felt that I did not need, nor did I want, his or anyone else's backing. I realized that it was important for me to acknowledge myself and follow what was really in my heart and what I felt compelled to produce. Why not? He found his niche and succeeded on his own.

In analyzing the motive of my long letter to Mr. Mack, I simply wanted to establish some kind of new dialog between him and me. I guess what I really wanted was his friendship, so that I could compare and grow from him. As an adolescent boy with Mr. Mack, I was always afraid to say what I thought. Now I wanted to go down to Miami and have a talk with him.

Well, it must have been about three weeks before I got a short two-paragraph letter, on a Burger King's Corporation letterhead, thanking me for my long letter and saying, "It looks like you have come a long way from the time you rang that dinner bell for me. I wish you success in your endeavors." That was the last communication I had with Mr. Mack.

Even though I attempted to later contact him by phone, I was never able to get any further than to his private secretary, Ms. Wheeler. She would always take my calls and would ask about my family; however, she always had some creative reason why I could not speak to Mr. Mack.

Eventually, I came to realize and accept the fact that Jim did not want to have anything to do with me anymore. I no longer satisfied the perception he held of me. Not getting the nursery meant that I would not be getting out of the army anytime soon. I had no regrets about writing that long letter to him; it allowed me to establish my direction and gave me my best confidence boost. From that point on, I had no fear of speaking up to anyone in the army. Frankly, I became very outspoken, no matter what their rank was.

Writing that letter gave me the chance to search deep within my psyche for what was important to me. Even though I did not have the understanding to go about doing anything about my inner findings, I knew that this mind thing really turned me on, and I wanted to go in that direction.

While in Texas, I entered junior college at night to study shorthand, music theory, and piano. My confidence had grown to a point of me being able to show some of my many quotations to the major in charge of the post newspaper. The editor liked my quotations so much he gave me a small column in the *Fort Hood*

*Post* weekly newspaper. I named my column "Can you dig it?" In the column, I would create little one-line philosophical statements for the soldiers to contemplate on. I would say things like, "Thoughts are the governor of one's reality," "Reality is the galvanization of beliefs," "The problem with problems are no problem once the problem is understood."

The editor agreed with me that no one was to know the author of "Can you dig it?" After these columns had appeared in a few issues, I asked one of the sergeants in the band his opinion on those articles. The sergeant said, "That is the most stupid thing I have ever seen." Man, was I glad he did not know that I was the author. Nevertheless, some of my most prominent and admiring readers were the post commander, a three-star general, and my division commander, a one-star general. The division commander was able to have his aid call over to the editor of the paper and, in fact, got my name. He wanted to have constant talks with me.

The general and I met on his open-door policy that he held once a week. We met and had a nice chat; he was sure that I had a degree in philosophy. I had to inform him that I had nothing but pure thoughts going for me at the time. We were both very comfortable during our wonderful forty-five-minute chats. I even told him about my interest in mind research.

The Brigadier General Smith was fascinated with my ideas. He also gave me a great confidence boost when he told me how my little column was stimulating wonderful conversation in his circle of colleagues. Finally, the main reason I wanted to come into the army were being realized.

I came into the army thinking that I wanted to compare myself with other men to see just how I stack. However, what I really wanted was to develop my self-confidence. Just to think that I was producing the catalyst for intellectual debates at the officers' club made me feel like Socrates.

# CHAPTER 8

## Thoughts

With my newfound confidence, I spent the next few years doing what I really wanted to do: research on how thoughts work. This was another SAQ, but it was the SAQ that drove me. I became less sociable as I turned inward to gain better insight on my thinking. It was necessary for me to spend a considerable amount of my life thinking about "thought" itself.

One of the first questions I considered was whether thoughts require an observer such as my consciousness or were thoughts the results of the observation of memories. Can thought be an observer of itself? I feel that one of our great breakthroughs in the area of thoughts will be a mechanical system that will allow the observation of thoughts. What a TV show that could be! And now broadcasting live from New York, a one-hour special into the mind of Chaz Cooper Disly.

You've heard the question, "If the tree falls in the woods, would it still make a sound if no one was there to hear it?" We know the physics of the falling tree dictates that the fifteen-pounds-per-square-inch average pressure would be greatly compacted the moment of impact (the tree hitting the ground). The compression of the air around the impact area would certainly set up a series of vibrations as air rushed away to make room for the fallen tree. The answer is yes; it would make a sound if the air were under pressure. The answer is "no sound" if there was no air pressure.

The question now becomes, "If there is no detector to detect this vibration, would it still make a sound?" If I do not have a working set of detectors (my ears), I could be totally oblivious of that vibration. Therefore, only if you had a functional set of air-pressure detectors would you perceive the sound of the falling tree in the woods. Such was the line of thinking I followed in tracing down the physics of thoughts.

The key element in tracing down and understanding the physics of thoughts is to first become aware and intimate with the "thought detector." For example, we have evolved or have been given (however one choose to view it) five detectors that enable us to experience the world outside of our body. If one was hearing impaired and no one told them any differently, they would go through life experiencing the world with just four other senses and never becoming aware of this fifth detector.

If someone came along later and made whatever correction needed to allow the experience of sound, they would still need training to be able to interpret the readings of this sound detector. Such is the case for our thought detector. What is the mechanism we use to detect and observe our thoughts?

Such a mechanism or thought detector should be viewed as another sense. A sense that had to develop out of enhanced needs just as our five recognized senses.

To better explain enhanced needs, imagine living a few million years ago when a hungry beast would sometimes sneak up on you and your loved ones and successfully devour one of your family members. It would not be long before you would have developed some enhanced needs for better hearing, as well as seeing. It is therefore through enhanced needs that such a survival mechanism came about over evolution.

Until now, we really did not need much more than those five senses to survive. We needed sight for locating things, so we therefore learned to see. We needed to hear danger when it was approaching us, so we learned to hear—as well as taste, smell, and feel. These senses all came about through enhanced needs and desire. This must be true for all animals as well.

In order for us to gain insight into the workings of our thoughts, we must first develop an enhanced need for the understanding of thoughts. Just look what we have been able to accomplish by Sir Isaac Newton's SAQs about "why an apple falls to the ground." Who would have imagined that from his SAQ we would be putting useful things into orbit?

As we bring clarity to our thought detector, our future accomplishments are unimaginable at this time, for such detectors will bring about an enhanced imagination far beyond our thought perception of today.

They say that a picture is worth a thousand words, but what is even more fantastic is, a thought is worth a thousand pictures. That is what it takes to have a thought—thousands of pictures flowing into our conscious mind.

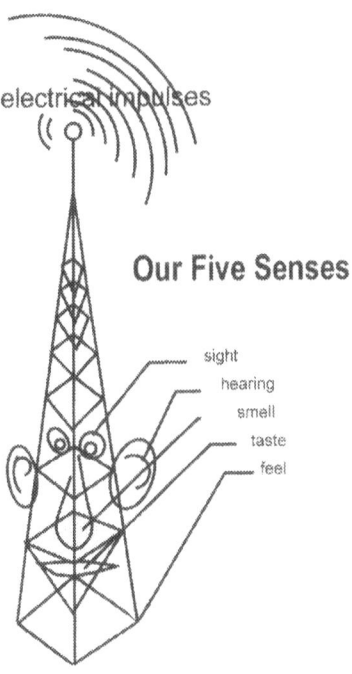

*Illustration of the Electrical Impulse from the Sense #8-1*

We sample the world with our five senses and then translate this information into electrical impulses. This electrical information is then stored as passive memory as it receives analysis through consciousness. Before that can take place, we should take a closer look at an impulse to see that a ray of information is revealed. As these electrical impulses are broadcasted throughout our memory, it can only be stored with other like memories already in storage. When it finds a community of like-thought memories, it celebrates with an illumination hue of familiarity.

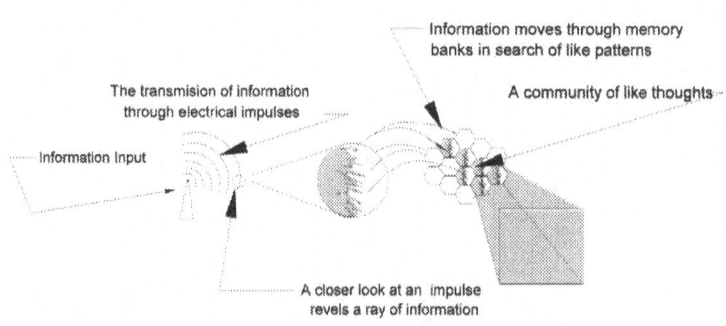

*Impulses to Memory*

To observe just one impulse illustration by itself would be unrecognizable and more like observing just one pixel of a picture, but the observation of thousands of these electrical impulses flowing in sequentially facilitates understanding as consciousness interprets. This then becomes more like observing a movie from the memory.

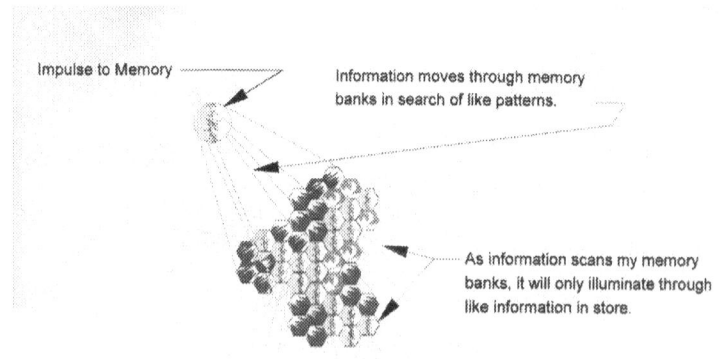

*Illumination of Like Memories*

When consciousness does not receive illumination for memory, these incoming impulses are nonrecognizable and a cause for alarm, as consciousness must give full attention when new cells are imprinted of the unfamiliar information.

With that in mind, one should then ask themselves, "Am I the observer of the consciousness, or am I the result of the incoming electrical impulse patterns?" What I am suggesting is that the impulse patterns flowing across special nerves could be creating an inner current which takes on a life of its own. Could this be what we are really referring to when we say "me"? Is my consciousness what I am referring to when I say "me"?

If we decide that that is what consciousness is, then we, the human race, can now move on and work with that as our hypothesis in that area. However, if we rather believe that we are the observer of our consciousness, then this would suggest and lead us in the thinking that we are a spirit experiencing life within this human body form. I find either of them to be very intriguing and absorbing.

The spirit idea would better explain the so-called out of body experience. Well, it does not really explain the outer-body experience, but it does allow me to isolate spirit from consciousness.

I have been fortunate enough to have first-hand knowledge of an out-of-body experience. Let me tell you, it made a believer out of me. It was like my spirit left my body with my consciousness as a passenger. I was suspended in the electromagnetic field, yet I was somehow attached to my memory banks. I had to be attached; how else could I recall the incident?

However, in considering what I call "me" as my consciousness, I (the me) would then be the result of the electrical impulses flowing into a CPU, a central processing unit, which actually thinks! This is not like the CPU of the computer that does not think. This CPU must be different. It processes the illumination of electrical impulses as the current flows through my passive memory banks. It is the passive memory that facilitates consciousness.

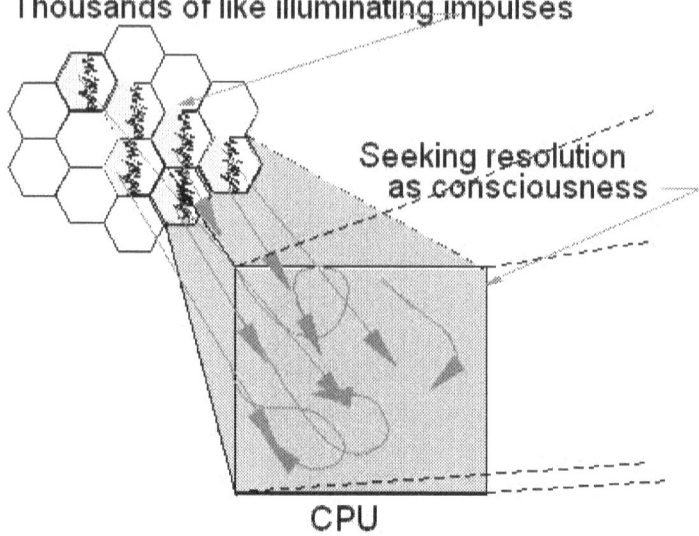

**Illustration of the Electrical Impulse**

These impulses seek resolution through illumination, but because thousands of impulses are streaming in, this illumination continues as it takes on a consciousness of its own. This consciousness comes about through a need to bring organization to the many random quantum particles being concentrated together. It is through this organizational need that a desire would come about to learn to control and respond to this incoming data of illuminations.

What brings about the organization and the control of this illuminating wave flow of hues traveling to its matching hue is consciousness—even though this consciousness only came about as a way of controlling the many hues flowing in to its like illumination.

A control of the incoming illumination takes place as the hue of the illumination is attracted into related hues. This hue attraction through discrimination creates the logic base for the hue organization. As the hues organize through a swirl of ongoing hue rainbow, the intense rainbow's activity within itself becomes what we call our thoughts.

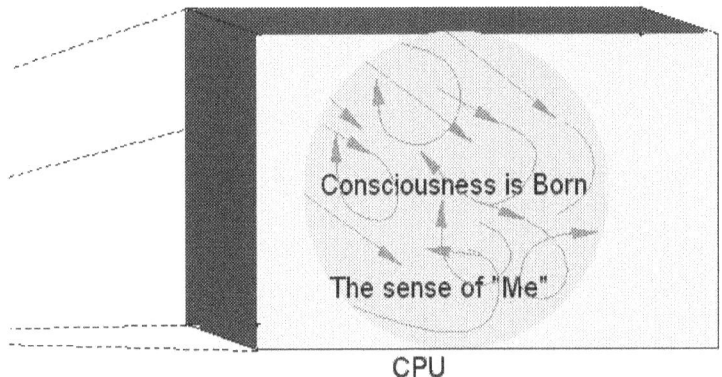

**Illumination into Consciousness**

Consciousness is therefore the illumination from the incoming electrical patterns as they resolve themselves in this now living light. The living light has learned to observe itself as it attempts to match each incoming hue with its conscious hues. It is through this hue activity that consciousness is born. This then is a CPU, a unit of consciousness.

Earlier in this chapter, I started out by asking whether thoughts require an observer such as my consciousness or thoughts were the result of the observation of memories. The answer is both. Consciousness came about because of memory, and yet memories came about because of consciousness.

By viewing consciousness simply as electrical impulse hues in a box, we can take a creative leap and move consciousness beyond the box. Beyond the box should be another book.

# CHAPTER 9

## Systems of Thoughts

Systems of thoughts are nothing more than one's routine or habitual way of thinking or performing a task. Following such systems, mini-operating programs, allow us some comfort zones. Without such systems, the world would be quite challenging. Imagine having to think your way through just standing up and walking. Thoughts systems become habits, and habits can become instincts. Here are some insights on some of my systems of thoughts.

Let us take information from one of my five senses—the system which facilities scent, that is. I sniff the air and gather information from that air's molecules. The characteristics of the molecules create a unique set of electrical impulses. These electrical impulses broadcasted throughout my memory banks can only line up with the prerecorded matching patterns.

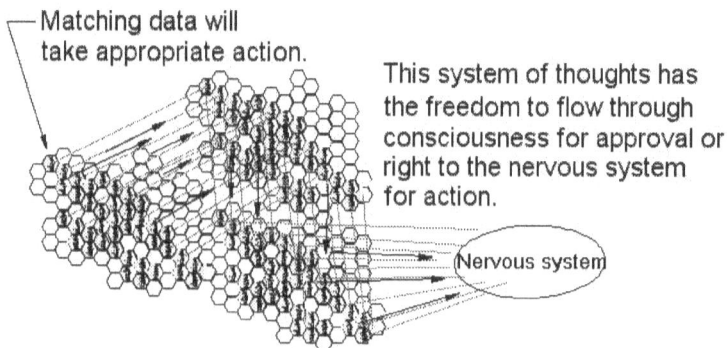

Matching data will take appropriate action.

This system of thoughts has the freedom to flow through consciousness for approval or right to the nervous system for action.

Nervous system

## A system of Thoughts

As the incoming electrical patterns ally with any prerecorded communities of like electrical patterns, they illuminate in a celebration of their union. This energy of illumination goes into my consciousness for interpretation. This interpretation is then sent back throughout my memories for any related stored interpretation, reaction, or any outcome of that reaction.

When a community of favorable, stored reactions identifies themselves, I can then follow the same procedural steps as that past system of thoughts. With these past communities of like electrical impulses located, also comes other stored data, like taste, sight, smell, and even the good or bad feelings about the information that fits those electrical patterns.

When a favorable reaction is stored in that system of thoughts, I am, my consciousness is, free to do other things. However, when the system requires modification, or if there were insufficient records or records of an unfavorable outcome, thoughts must be given as to how I should proceed. A person with a damaged memory may have no way to employ their system of thoughts and, therefore, would need to think their way through the simplest task. Even the tasks of understanding would pose a challenge when there is no access to the stored related patterns to the conclusion of, perhaps, the air molecule event.

When faced with an unfamiliar situation, like no system of thoughts, I may just have to rely on my emergency system of thought. Run like hell. Do not laugh; I did not create that system; I inherited it from my many past generations.

Apparently running could be, and most likely was, a wise system of thought to engage when something was unfamiliar. Why stick around for possible destruction? So, now that I have chosen to run, my run system of thoughts and the procedure for running have employed automatically.

Perhaps some are thinking that running does not require thoughts. Well if you are thinking that, you are right because I have already stored the thought system for running years ago when I learned to walk. Once my fear factor sets in to flee, my nervous system only needs me to say go. A stored set of memories has everything all worked out for the run.

On the other hand, if I find an odor to be unknown, yet my eyes did not confirm any danger, I may stay around to search out more data like, "What is that sound? What is that taste? What is that smell?" If my other senses can confirm a safe experience, I may just stick around.

Like my example on smell, sound works the same way. It too comes in as a sampling of the air molecules. This time, however, I employ my atmospheric barometer instrument, better known as my ear, to measure the slight changes in the pressure of the air molecules. Vibrations from the changes in air pressure transpose into impulses of electrical information patterns. This impulse resonates throughout my memory banks, as it reaches my consciousness, and is understood as a particular recognized sound.

When an incoming sound pattern lines up and is accepted as a matching pattern stored as sound, the electrical impulses line up and pass through the like patterns. As familiar electric impulses proceed to its primary community group of memories, this group transmits all of its stored information about the molecule structure of the vibrating air pressure into consciousness by way of illumination.

Another discovery I have made in dealing with my systems of thoughts is that my consciousness can handle only one thing at a time. When I smell, I am not seeing. When I am hearing, I am not feeling, etc. I have learned that I view information in vibration sequences, and it happens quite fast, so fast that it appears as if I am doing it all at once. With the illumination of so many quantum particles, it takes on a consciousness as a way of organizing and understanding itself. With this taking place at the speed of light, it is no wonder many of us may not have been able to work out this function.

Perhaps you have noticed that when a person is straining to hear something, they have the tendency of closing their eyes; this is to shut off the sight sensory. Try turning off the lights in a dark room. You will hear so much better with the sight sensory no longer being observed in the vibration flux. Consciousness focuses only on one sense and one thought at a time. With the sight sensory closed off, hearing becomes enhanced.

We do have the ability to focus on just one sense and heighten that sense tenfold, if not more. I remember how my mom could sometimes hear what was said when she was in another room. I used to wonder, "How did she hear that?" My mom was able to focus on her sense of hearing by remaining still and half shutting her eyes, obligating her total consciousness to the sound vibrations. Not just any sound, but the sound she chooses to isolate and focus on.

That ability to isolate and focus on just one set of vibrations is nothing short of incredible. To sit in a concert and isolate the vibration of any single instrument performing proves to us our natural abilities.

Being aware of this astute ability should enable us to recognize that we are capable of cutting through any chaos and focus directly on whatever we choose. We need to have a clear understanding that our focus controls our consciousness by steering it to what it wants. However, it is our consciousness that establishes our focus. This sounds like double talk, but it is not. We will work through this paradox in chapter 16 as we deal with consciousness.

Growing up, we had to rely on our parents to make decisions for us and establish our systems of thoughts. Yet, as adults, many of us, out of habit, still look to others for their systems of thoughts. Looking to others is in itself a system of thoughts. I mention this because one can never reach their spiritual potential when they continue to rely on others for their system of thoughts. No one outside of yourself knows your spiritual needs.

If we are to ever develop our minds to the state we had before the Great Flood, then we, as the whole human species, must first be able to evaluate and make some paradigm shifts for ourselves. One must, first and above all, have freedom of their thoughts. One's mind cannot entertain creative thoughts when shackled with antiquated thought systems.

Imagine what our civilization would be like today had we continued allowing people to convince us that the earth was the center of the universe. Science would not have advanced if it were shackled by inflexible antiquated thoughts.

One of my antiquated thoughts taught to me was that God destroyed the earth by water because man was too corrupt. Of course, I now know that the earth is constantly shifting as it evolves, and the best thing we can do is to learn and understand our earth, so that we can stay out of harm's way.

While there is great wisdom in the Bible, any unshackled mind can see that many parts of the bible were translated out of pure bias and ignorance. One can find overwhelming biases toward women for example in most of these religious texts. It is all justified as ( )'s will and wishes. As I take notice of the academic accomplishments of my daughters, as truth seekers I see self-sufficient human beings capable of taking care of themselves as well as their families. There is no doubt in my mind about the capability of women. I am sure you would agree that anyone trying to pigeonhole women or any other group into a subservant role displays their vulgar ignorance.

Today's society dictates that each individual be self-reliant as well as free thinkers. I think that it would be great if everyone on earth were encouraged and allowed to be all that they can. Anything short of that is wasted time of one's spiritual development on earth.

It should be a crime against humanity for anyone to enslave or hinder the growth of another human being. Even when that individual is not aware of their rights as a human being, the spiritual need for freedom is essential. We can never know which mind will contribute what to humanity. There should be no exceptions, not even marriage. Marriage is no reason for the domination or suppression of another person's thoughts and spirit. Why create that kind of memory in someone? Memories are the food for our thoughts; our thoughts are the essences of our consciousness, and our consciousness is the quintessence of what we call our spirit.

# CHAPTER 10

## Spiritual Gratification

One day, I was speaking with a minister friend of mine when I mentioned the phrase "spiritual gratification." My friend hopped on the phrase as if he was an authority on the subject. He was so eager to display his spiritual understanding; I could not get a word in. He never gave me a chance to explain.

What I meant when I said "spiritual gratification" was to satisfy the direction a person's spirit wants to lead them. If we are to accept the idea that some spirit wants to lead us, then what and where does this spirit come from? I say it comes from that black hole in the center of the universe. Permeating from this black hole is a pulsating spirit that is pulling on everything.

We, as well as all that is in our universe, are the results of this pulsating source. When we refer to our spirit, we are referring to the unadulterated magnetism that encompasses the whole universe. This spirit is a *pulsating magnetic source* which I will call PMS.

PMS, unbiased by nature, permeates throughout our consciousness, as well as every form in the universe. PMS emanates from the primary universal black hole in an attempt to pull itself back together again. Through this recoiling attempt by PMS, the whole universe went into an internal orbit. This orbit or swirling action caused the neighboring particles to clump together, creating all bodies within the universe. We get a sense of this phenomenon in what we understand and experience as gravity.

We can further imply that PMS is the igniter of our consciousness and can be better discussed at this time if we recognize it as our spirit. PMS are not only what we call our spirit; PMS is the spirit of all that is. Even though PMS permeates all that is, we still have the freedom of our will, which would then put PMS more as the observer of our consciousness.

It is through PMS that we, as well as all things that is, receive and connect to the Creator. Through each of our prospective, the Creator can comprehend self. The functioning of PMS may have been so incredibly simple; it has gone virtually unobserved until now. Incredible as it may seem, PMS is so common; we may have been just too close-up and intimate to even recognize it.

In this chapter, we can give thought to a small facet of PMS as it pertains to us directly.

To think that our every thought is being observed through PMS is like discovering that a Trojan horse has been operating on our computer hard drive without our knowledge; this would certainly cause some feeling of violation. However, in the case of PMS, this should not be the case. PMS were the initiator, the operator, the instigator, as well as the stimulator of our every thought in the first place.

How does one really know that the thoughts they are thinking are even theirs and not the thoughts of others by way of PMS?

If PMS is of the Creator and from the Creator, then whose thoughts are they anyway? Without PMS, there would be no thoughts at all. There would not be any you or me.

Receiving thoughts through PMS does not necessarily suggest that any of our thoughts are from the Creator. However, it does suggest that with everyone's thoughts being taken in through PMS, a universal consciousness is formed as these ideas seek their vibrations of like hues. As like hues corroborate, so does the universal consciousness of humanity evolves. Get enough people thinking badly; the world becomes bad. Get enough people thinking hate; we become hateful. Get enough people thinking greed; the universal consciousness will produce gluttony, voracity, ravenousness, insatiability, hunger, self-indulgence, and even a government that will embrace these ideologies.

When we become more aware of our consciousness as it relates to PMS, we must learn to think good thoughts if we are to survive as a species. If you do not believe me, just keep allowing people to think war and see what the universal consciousness will bring about. Better yet, do this little experiment: observe any group who constantly think about getting into fights, and see how often a fight manifests itself in the life of that group or an individual.

Obviously, I equate good thoughts as those thoughts that enhance humanity, and bad thoughts as those that will eventually cause the demise of humanity. Therefore, when I reference thoughts as good or bad, I am addressing the outcome of that thought as it affects our universal consciousness.

The phrase "spiritual gratification" could also suggest the idea that our spirit enters our bodies from birth with some preconceived agenda. Not so, I think that spirit or this PMS causes a different effect on each individual body due to the particles that went into the making of each of us, the makeup of our consciousness.

This PMS function is the catalyst that caused the swirling of particle into the forming of stars and planets. Stars and planets all have different characteristics because of the difference in the particles that went into their formation.

Each of us is also different, but certainly not because we have different PMS (spirits); we are all influenced by the same PMS. What makes us different is the uniqueness of the particles that went into our personal makeup. PMS can, therefore, be described as a spirit, because it is what influences any particular bodies.

Another point of view into PMS comes about when we are perplexed with a problem. We usually turn it over to our subconscious to find the answer. Our subconscious, being unrestricted by the memories of logic, right or wrong, differ from consciousness in that it can imprint onto newly developed brain cells which have no particular patterns.

### New brain imprints

### Sub-conscious at work

Once these newly imprinted cells have had the opportunity to develop into a community of like cells, these communities then illuminate into even greater hues of subconsciousness. The subconscious hues imprint back and forth from consciousness to new memory cells until the community is sufficient to support a resolution to what was once a perplexity.

Such communities will sit dormant until the conscious mind calls on it. Accepting this idea as a normal universal function in all animals then leads us to the supposition that ( ) satisfies its understanding of itself the same way, with each of us as the equivalent of one of ( )'s brain cells.

Imagine the accomplishment of one's total life span as the hue file of just one brain cell of ( ). As individual brain cells of ( ), we are endowed with all the nourishment needed to work out any spiritual complexity that came associated during our birth. The situations needed to bring about the solutions to any spiritual needs were born within each of us at birth. That is why some people go through life attracting situations, based upon some prime objective, for which they have no apparent conscious reason.

Now with this awareness, one should not be in despair with the appearance of their reality at any given time. We should become sensitive to the spiritual needs within us and find a way to gratify that need. This is the only way we will ever experience true happiness or spiritual gratification.

With the PMS serving also as an information medium—training one's consciousness to surf this ubiquitous PMS—the gratification of any perplexity would be joyfully at hand. Without this realization to face life, one could experience a sense of hopelessness and never fulfill their spiritual quest.

Our conjecture thus far is that PMS is ubiquitous and extends from the black hole as a wave function, which by the very nature of the function has created all that is. I further hypothesize that spirit, as it exists in us, serves as the stimulus for our consciousness as well as an observer of our consciousness. This then must lead to the idea that through us, our Creator understands and perceives our reality as one of its universal perceptions of itself. This should then bring us to the question, "Whose thoughts are we thinking anyway?"

How does our individual spirit come about, and how does it work? What is the relationship between PMS and our spirit, if any? Let us take these questions in the reverse order, starting with PMS and then spirit. We learned that PMS is the ubiquitous pulsating magnetism source, capable of manifesting any vibration wave particles into its preconcluded reality

On the other hand, we can view our personal spirit as a surfer of PMS, free to surf throughout the waves of PMS, unrestricted for hues of familiarity.

If I have been skillful in presenting the previous ideas, you should now be ready to take the upcoming conjecture leap with me. Since thoughts can only congregate within like thoughts or like hues, it should be easily recognized that such thoughts could be of hate, love, war, law, order, you name it, and it is out there as the consciousness of the universe.

As these various hues randomly surf the PMS for like hues as a possibility for resolution, they sometimes find their way into the ignition of the consciousness of a new mind being formed for the first time (the birth of a baby). The hue that ignited the new mind into consciousness is within itself a small hue facet of the universal consciousness. Once isolated within a human being, we call it the human spirit of that person.

This spirit or hue, now cased within the newly born, will still seek the congregation of like hues to justify and understand itself. Whatever type of hue that happened to ignite the consciousness will reflect this newborn's prime objective or its purpose in life.

With that understanding behind us, there is no need to dislike someone for the way their spirit motivates them. It appears that we are who we are regardless of what someone tries to make us out to be. Alternatively, we can say that we are the spirit that drives our action. An interesting question about this arrangement is, "Who is the boss?" It was by the spirit of the hue that ignited our consciousness into existence, yet it is through our conscious will that we direct our lives.

We all know that when spirit decides to leave our bodies, our consciousness can no longer exist. This would put spirit at an advantage over our consciousness. With that in mind, we could concede that spirit is the boss.

Therefore, if the spirit is the boss and all we are really doing is working for the spirit, it is not so much of a leap to land on the idea that if we do not try to gratify our spirit, we could wind up with sickness and disease, because we are pulling in a different direction from which spirit leads.

This chapter should have been a lunch basket of scrumptious thought morsels for you to feast on for some time to come. Remember to take your time as you delight your consciousness with these delectable ideas in each thought bite. I hope that you are enjoying my thoughts in this book, as much as I am enjoying writing this to you, my wonderful family members.

# CHAPTER 11

## Focus

Focus is the most crucial mechanism of our intellectual endeavor for spiritual gratification. Without a strong focus, it is difficult to steer our consciousness into a particular direction. Even if one is fortunate enough to discover the nature of their spirit, it is not possible to succeed in that direction without the ability to stay focused. When taking on any extended task, focus is essential for a favorable outcome.

As we concentrate on focus, we can imagine focus as a small light always on in the corner of our bedroom. Whenever we wake up, a light in that corner of the room is always illuminating a primary picture we have placed there. The illumination from the picture in that corner of the room is the first thing to capture our consciousness each day.

As incoming data illuminates into our consciousness throughout the day, consciousness will remain faithful to its primary picture hue. Any incoming hues can only be of service to the governing picture hue, in other words, our focus. Even after we leave our imaginary bedroom, this primary hue has established itself as the governing hue thought force for the day.

Without this corner light on to keep our hue focused, we would wake up following whatever stimuli happened to capture our imagination. Such a focusless mind would succumb to whatever interesting thought happens to come along, like some aimless teen following whatever feels good at any given time.

As for the mechanical functioning of focus within our minds, it starts out with a well-defined system of thoughts stored in the brain cells as memories. Memories that are constantly illuminated will hold a bit of its charge of illumination. Frequently observing the goal center, the picture on the imaginary wall, will keep the hue charged in the illuminating consciousness process.

This night light becomes the primary point of focus, causing all other stimuli to flow in and out of consciousness as just a passing thought. These thoughts can only be fleeing thoughts, because the night light focus controls the direction of consciousness. Only hues of like thoughts can harbor in this particular hue consciousness.

Any person not having a hue picture to focus on will experience an inability to stay goal oriented. Whatever thought comes to mind is pursued due to their lack of a picture in hue. Such a person would appear docile because of this inability to hold any objective in consciousness.

Even the task of telling a story could prove difficult without a focusing ability. The story being told by the storyteller would stimulate other nonrelated thought hues into their consciousness, overwhelming their ongoing hue consciousness and causing them to lose touch with the very thought they were pursuing. The listeners are led up some rabbit trail, which may have little, if anything at all, to do with the storyteller's original tale. Now, the listeners, as well as the storyteller, start to wonder where this story was going.

For a person experiencing such a problem, allow me to create some software designed to eliminate it. First, give any story you wish to tell a short title in your mind. The title you choose for your story should be one that you will find necessary to say throughout your story. This title then becomes your point of focus to keep your consciousness locked on your story.

Learning to give a mental title to your stories before you tell them will afford you the successful telling of your stories. Your proficiency with this little software technique will escalate with practice.

On the other hand, there are those with such a deep focus that they lose track of time, forgetting to eat and sleep. We all know someone to some degree of that state. They often display some idiosyncrasy, which normal folk finds strange.

However, it is only when a person learns to condition themselves to deep focusing can they follow their path. It is crucial for everyone wishing to accomplish his or her spiritual gratification to first look within and follow the hue on his or her yellow brick road to whatever wizardry they are to produce.

# CHAPTER 12

## The Danger Involved

I pointed out in the introduction that the misuse of this material can be detrimental to you as well as your loved ones. Let us start with the phenomena known as stigmata. Stigmata is a phenomenon that some very religious individuals experience through the bleeding of their body. This bleeding occurs through mentally self-inflicted wounds that they have been taught Jesus received due to the person's extreme feelings of sympathy. Some stigmatisms occur in the palm of their hand, while others occur in their wrists or forehead. Where the wounds manifest depends purely on which artist they have been preexposed to. This is a very strong hint that we have the ability to destroy or heal our bodies with our minds.

Such phenomenon could better be described as an altruistic psycho manifestation.

Such phenomena are said to be a miracle, and that is true. There is a miracle. The miracle is that we have been given a hint into our true abilities. Thoughts should be given to enhance this ability.

Another phenomenon that some still observe is that of superstition. This is a misleading notion, and we are all familiar with someone who still practices this destructive way of thinking. Everyone needs to know by now that it is our beliefs that govern our reality. If one person thinks that they will have seven years of bad luck for breaking a mirror, then they will. They will because it is their idea of bad luck that they themselves manipulated into reality.

Another interesting phenomenon is the people who receive scratches on parts of their body in a belief that a ghost attacked them. Again, like the blood, the scratches are real. But the ghost is nothing more than their subconscious manifesting their belief on their body. A better description for this phenomenon should be self-abusive psycho manifestation. Our medical society should take heed at this phenomenon. If controlled, this possesses great healing possibilities as well.

What we are to realize about such phenomenon is that we all have this ability, not only for those manifestations within our bodies. We also have the ability to project and manifest outside of our bodies as well. The so-called sightings of ghost are nothing more than quantum energy being manifested from the subconscious mind as thought waves are morphed into its counter particles.

Such manifestations can be quite dangerous when one does not understand their mind power. One can only imagine the many people who have frightened themselves to their death. That's why we must learn to "fear not." The best reason to fear not is because we manifest our reality through our beliefs. Those thoughts of fear can only manifest itself if that is what we hold in our mind as part of our focus. The spending of one's time on what they do not want to happen should never occur. Continue in your imagination if you need to organize and eliminate this thought process.

However, because the universe is balance seeking by its nature, every thought is also confronted with an opposing one. These, then, become realities of conflicting thoughts as each community of thoughts competes for control of the universal consciousness waveform.

Religious leaders recognize this as spiritual warfare. No matter what one may call it, we can and should stop this deceitfulness to ( ) as each of us take it upon ourselves not to lie, cheat, and deceive others no matter how much money we can make. As we learn to acknowledge our quantum connection, we will have no reason to cheat the universe of its understanding.

We first teach our children the art of lying with old St. Nick, and now we sit and wonder how these children, now CEOs, could be so corrupt and insensitive to workers. What we are really seeing now is a St. Nick syndrome. Nevertheless, are not the CEOs' parents really the abuser? Should not the parents of those CEOs be put in jail as well and receive punishment right alongside their St. Nick kids? After all, when parents are found guilty of child abuse, they are punished. Anyone instilling the concept that "through conspiracy comes gifts in abundance" abuses and corrupts a child's mind.

Nevertheless, corruption must never go unchallenged. Even if we adversely teach corruption to our children by deceiving them each Christmas, all of humanity will receive the punishment because that corruption can only be reflected exponentially throughout the universe. How long will it be before all of humanity starts receiving the feeling that everyone is a thief? Is this the world we want to leave to our children?

If you are familiar with the hundredth-monkey theory, then I would say that we must be fast approaching the ninety-ninth monkey. It would delight me if people could realize that no amount of money could secure them or their offspring as long as corruption permeates the universal consciousness.

Earlier I stated that our thoughts are attracted by like thoughts. Spirit works somewhat like thoughts, in that it can only attach itself to like spirits. When a spirit leaves the body through the out-of body experience or death, the spirit (because of its frequencies) gravitates to like spirits. Therefore, such a spirit can only accommodate one who has fed nastiness into their spirit. To one who has been sweet, sweetness is only where this spirit can reside. It works the same for the spirit of those who are deceitful.

I was once in business with a guy who was a liar and a cheat, such was his nature. Ironically, he wanted no part of other dishonest people. He only wanted to surround himself with sweet, honest people. His spirit is going to be in for a big shock when it finds itself with like spirits.

The person I referred to takes advantage of everyone who has the misfortune in dealing with him. I too shared in this misfortune, but I have learned that it is better to leave the quantum connection to deal with them. The universe has a definite resolution for all things, because all things are of the black hole's function.

The grip is through the electromagnetic force that permeates throughout the universe, encompassing all that is in it. Our thoughts and our spirits, like the planets, are also of the universe and are influenced by the electromagnetic phenomenon from the big black hole as well.

Because electromagnetism abides by very strict rules, anything contrary to its laws must cause an equally strong reaction. The universe seeks balance by pulling everything back in; anything out of balance is rejected. Though it might appear as if one is getting away with their deceitfulness, they are just fooling themselves. A spirit can only reside within its frequency, the frequency in which the person created.

The sad thing about all of this is that when people are unfortunate enough to have been bought into the deceitfulness and negativities, they will suffer because of the universal negativity. No matter how convincing the skilled negative advocator may be, the universe can only react by ushering in this negative reality, a reality that is only met with inevitable violence. The greater the negative input, the greater the negative response from the universe. Adolph Hitler, one of humanity's negative advocators through oratory skills, created a reality that opposed the universe's electromagnetic laws. Such is the case when operating from a platform of universal ignorance.

Have you ever noticed how people start to get religious when they realize their end is nearing? Many sense that they have a spirit that could leave their body and go elsewhere, by the electromagnetic field perhaps. Our whole life is spent building our spirit, and yet many of us never quite make contact with it. It seems ludicrous for one to think that what they do in the late winter of their life is going to unweave their summer's indiscretions.

I guess in view of the like-thought theory however, any changes one makes will most likely facilitate a different reception location for their spirit. They could fit in with all the other deceitful spirits like themselves who made a change of heart. That change is ours to make. The universe has room for all our spirits to settle into.

The true danger in manipulating quantum energy out of ignorance is that whatever we hold in our hearts will come to pass— that is, as long as we hold on to the idea of wars as a solution, wars will be manifested. If one allows ignorant people to convince them of, for example, an Armageddon, then an Armageddon is what will manifest. Whenever we allow people to convince us to ally in their negative beliefs, we are then the cocreator of that negative reality as well.

You should have read enough of this book to now know that you are the creator of your universe. You are at the center of your universe. Being the center of your universe does not take away from anyone else's center. No, the world does not revolve around you; your whole universe does. With that in mind, take care in the reality you create and avoid the danger of negativity.

# CHAPTER 13

## Religion and Its Contributions

My prayer to ( ) was, "Let me find the truth about the universe and, most of all, about you." I knew that if my prayers were to be answered, I would have to keep an open mind about what would be revealed to me about ( ), especially if it is in direct conflict with what I have been taught to believe. That became somewhat frightening as I did work in this area. I knew that if I started inquiring about my intrinsic beliefs, I should have a good substitute in order not to experience panic attacks.

As a child of five, I remember Santa bringing us a roomful of wonderful toys each Christmas. We were very poor, so the only sure way I could expect toys were from Santa. When I went to the first grade, my classmates made me cry when they tried to convince me that there was no such thing. They wanted me to believe that my parents provided all those gifts.

As Christmas grew near that year, I started searching around the house for hidden gifts as my classmates suggested. To my disappointment, I started locating many hidden gifts. This was my first hint that my parents would lie to me. That was not a good feeling. Even worse, I had to finally accept the reality that Santa was an elaborately constructed lie.

I felt that I could not talk to my parents about my discovery; I dare not entertain the thought of them as liars. After that Christmas, I quit asking for any gifts from Santa. My parents thought I did not want to ask Santa for any gifts because I knew that I had been a bad boy.

The loss of Santa made me feel unhappy and less secure. The realization that my parents went through so much trouble to provide the many gifts was overshadowed by my perception of their conspiracy. Life was never quite the same for me after that Christmas. This was the first time I could recall having that nervous feeling, which I later recognized as panic attacks. I later experienced it when I started questioning my religion.

For years, I had real problems with the idea of the whole congregation tithing 10 percent of their meager income as the pastor lives like a king from our effort. His teaching on reciprocity never made sense to me, until I discovered the quantum connection and how it related to the universal consciousness.

I have never been in a church for any time where the pastor did not turn out to be a sneaky wrongdoer. Whenever they are exposed, they say things like, "Keep your eyes on ( ) and not on man." Just once, I would love to meet a truly honest pastor.

With the spread of television and the Internet throughout the planet, many religious leaders are exposed a lot more readily. Yet, with their exposure, we are still willing to forgive. Why is that? I could not help from noticing that when the pastor wanted more money, the congregation was bombarded with messages on reciprocation.

Before you shoot down on me as being sinister, allow me to share just a couple of my many church hurts: Some years ago one of my most admired television ministers had convinced me that, as a single man, I should practice chastity. To the annoyance of my female acquaintances, I practiced it with pride. I was devastated when this minister was exposed for his adulterous prostitution-patronizing habits.

Another deep hurt occurred when I was part of a small congregation of about thirty-five members. I joined their worship team of a piano player and six vocalists and became their instrumentalist, embellishing their voices with my trumpet, harmonica, soprano, or alto sax. Within eighteen months, our congregation had grown to over three hundred tithing members.

One of the things I enjoyed most was being able to quickly find the key the pastor would start singing in after finishing his message; what a beautiful voice he had.

So what could go wrong with such a wonderful, prosperous ministry? The pastor, an educated man with three smart kids, had to leave the church and skip town to get a divorce, so that he will be able to marry one of the college youth of the congregation he had impregnated! I could never understand how he could have left his sweet, charming, educated wife.

I could see why he taught so much on forgiveness. He knew what was to come. I left the church soon after, not because I could not forgive him. It was because I was so hurt to think that no one thought enough of me to tell me what was going on. I am sure, had someone informed me earlier, I could have made a difference.

These are the things that make me wonder about the hidden agenda behind the message. I have also come to believe that if religious leaders are not able to find new ways to acknowledge truth, especially when it conflicts with the model they have been teaching from, religion will most likely not survive this millennium.

Religious teachers must keep abreast with science, so that their messages will make sense with the irrefutable scientific discoveries that are made each day, like DNA and genetic engineering. A rule that should be kept in mind is, **"There can be no confusion in science and** ( )," because ( ) is science and science is ( ). The confusion, therefore, must be in our model of ( ).

A good model will fit every situation and will work for the past, present, and future. A model does not have to be perfect to work. However, as our understanding precedes the model in accuracy, changes must be made to the model, or a new one must be found.

There is no doubt that religion has brought continuity to our human experience. It is society's attempt in creating and instilling some sort of equitable rules into our consciousness— a set of rules that could apply to any given situation we may find ourselves in. Religion has fashioned the foundation upon which our faith rests.

What is most perplexing about the framers of most religions is the way they fell short in dealing with their intolerance to other religious persuasions. These religious institutions were all set up by the thinkers of their time; they were the visionaries of their day. Many of them saw the need to even break away from the dogma of the status quo and start new religions.

By encompassing the old beliefs, they embellished new dogma that later became perplexing to its future congregation. Let's face it, if they had gotten it right, we would not be killing one another over our present religious beliefs!

Killing each other is like a cancer among the human population, the brain cells of ( ). Try looking at each human being as one of ( )'s many brain cells that were given birth for the purpose of solving and understanding the universe. None of us knows what the next brain cell will and can solve, no matter how lousy they may appear to us. What we should all know and acknowledge is that each of us, no matter what our bias tries to tell us, have been given birth through the desire and needs of a much higher consciousness.

Consider this: As creatures ventured onto land from the sea, their fins that evolved for swimming evolved into limbs out of the necessity to move on land. As they eventually started standing upright to see better, a stronger set of rear limbs came about in support of the total body weight. This freed the front limbs to evolve into what we call hands. This whole evolution came about through ( )'s desire to perceive its universal self. Through this process came the evolution of our brain that we are still learning and relearning to use.

Each of us is the way in which ( ) has chosen to explore its universe. Well, why not? Shouldn't ( ) be able to do that in any way it chooses.

So, what was a religious composer to do when all he had at his disposal at the time was the philosophy of his preceding framers, an-eye-for-an-eye concept, his knowledge of the universe as he perceived it to be, and his discontent with the status quo? When inputs of distorted thoughts fill the universe, we can only expect a distorted output.

Distortion in, distortion out. We can only input and draw on what we thought to be the truth at the time. However, we do not have to get it 100 percent right the first time, but we must be trying and going in the right direction. For example, our initial modeling and understanding of the atom was not quite right at first, yet we went on into the splitting of the atom.

It is okay to proceed with some distorted knowledge of the facts, as long as we are willing to adjust when new understanding of the universe unveils itself. Classic Greek thinkers had us westerners thinking that the earth was the center of the entire universe. The facts were somewhat distorted, but it was okay. It was okay because thinkers such as Aristotle tried to make sense of things not understood at the time.

Then came the Polish astronomer, Copernicus, who in 1543 published the first contradictory facts about the immovable earth as being the center of the universe. The keepers of the problem met this with bitter opposition. Because the keepers of the problem had control of the mass' minds, they were able to discredit Copernicus' publication. This suppressed the creativity of individuals for years to come.

Ninety-one years later in 1632, Galileo, another well-known astronomer and a seeker of truth, was compelled to make public his findings about the planets. His observations using the telescope proved that the earth was not the center of the universe, but only another planet orbiting the sun with its other known planets.

Unfortunately, for Galileo, the religious leaders felt so threatened by his findings that they had him placed under house arrest for the rest of his life. They were not willing to say that new information had presented itself for the advancement of humanity, and they could not acknowledge or confirm these findings at this time. Like little kids with power, they felt threatened. They had the power, which hindered the creative minds of humanity for another two centuries.

One cannot help wondering why they were so threatened. Was it because they felt that civilization would fall apart due to some diversity in thoughts, or was it something more selfish and sinister, like a fear of the mislaying of their personal placemats?

One thing we can surmise is that Western Civilization was greatly influenced by Gothenburg's Bibles illustrating elaborate drawings of the earth and all the heavenly bodies revolving around it. It took some two hundred years after Galileo's publication for some solvers of the problem to withdraw those distorted models from the Bible. Finally, the minds of truth-seeking students were free to think in a more realistic way about orbiting bodies, gravity, mass, speed, force, etc.

I hope that with the creation of the Internet, the keepers of the problems will no longer suppress information so easily. The Internet has made the business table a lot more level for anyone wishing to communicate their ideas with others throughout the world. If we are the part of the instruments in which ( ) perceives its reality, then we must learn to be tolerant with the diversities of other's thoughts and ideas.

Personally, I find the Bible to be filled with great wisdom and, at the same time, a great deal of distortions. It is up to the individual to see distortion for themselves. If one does not see, then it may not be for them to see, because we all serve the consciousness of ( ) in different ways; our purposes will therefore be diversified.

Here is something to give thought to: the Bible states that revenge should not be ours, but that of the Lord. There is a saying that if someone does an injustice, we should not strike back with another injustice. This certainly would suggest that the battle is not to be.

Well, that is a pretty tall order. Just how does one avert the pain and anger brought on by this wrongdoer? Pain and anger usually trigger an overt retaliatory action.

The only way I can see such a philosophy working is for one to have an unconditional feeling of connection and love to that wrongdoer, as well as to all human beings. You would most likely have to understand that the wrongdoer is acting out of his or her ignorance of their quantum connection. One would have to know that this wrongdoer is operating on a primitive, unenlightened state of mind.

If a snake or an alligator struck us, you would not see the animal as a wrongdoer, because you know that such an animal operates in a more primitive state than you. You would most likely overlook the animal's action with forgiveness.

Why then retaliate against an unenlightened human animal? Just because the animal is human does not mean that they are enlightened. If you pay attention, you can find unenlightened people in some of the most powerful places, and still they are not aware of their human connection other than what is in it for them.

What you would have to know about the unenlightened wrongdoer is that they would love you as you love them once they make their quantum connection. Getting a copy of *The Quantum Connection Theory* in their hands would be the most far-reaching action you could possibly take. We know now that we cannot change someone, but we can give him or her the tools that would allow their enlightenment.

Even so, what should be the procedure if it becomes clear that this unenlightened wrongdoer's act was intentional? With the understanding of the quantum connection, you go to ( ) as in meditation or prayer, acknowledging that you have been mistreated by the said wrongdoer. You then turn it over to ( ) and just let go. I have done this many times in the past, and I have been astonished of the results. You will never know until you have tried it for yourself. I will tell you this: I would have never dealt so harshly with my wrongdoers as ( ) has.

Another very interesting thing will start to happen, and you may not even realize it. Wrongdoers will diminish, if not vanish, from your reality completely. This event can no longer take place, because you no longer hold such a reality as part of your consciousness. Before long, our quantum connection will surpass its critical mass as you do your part by releasing and showing love and understanding to wrongdoers. This has nothing to do with what is right or wrong; this has to do with what we hold in our consciousness.

# CHAPTER 14

## Accepting the Connection

In this chapter, I ask that you extend to me your most dilated imagination. Accepting the connection must be personal. No one having the need to seek another person's opinion about their own connection to the universe has matured enough to know that other people can only respond out of their own mental cave.

One should work out within themselves, as they learn to accept their quantum connection, how willing they are to examine their beliefs, to change any negative thoughts to the betterment of humanity, to trade in any belief found that could cause the demise of another human being, and then ask themselves, "Is such a connection to ( ) worth it all?"

As a consciously connected individual, one should establish their own sense of peace and harmony, while knowing that their sense of peace and harmony does not violate the peace and harmony of anyone else. A connected individual accepts whatever he or she needs from the universe in great abundance without hoarding simply by keeping it in circulation. Once a person understands their quantum connection, there is no longer anything to fear.

There is nothing wrong with anyone making lots of money. The accumulation of wealth through some goods or service one has provided is a great yardstick of their accomplishments. Everyone should be handsomely rewarded for their contribution to society. This is a wholesome and wonderful way to promote the highest achievements from all individuals. To earn wealth is a lot different from hoarding or stealing for wealth.

Human beings are all wired up for the quantum connection already. Unfortunately, if we have not discovered, or not been taught, that we are already wired up, we could spend our whole lifetime blundering and trying to make a living just to get through life. Jesus once said, "These things you see me do, you will do and more" (John 14:12). Was Jesus just blowing smoke, or is it just that we think so little of ourselves to believe that we can do the "more" of which he spoke of? Perhaps now is a good time to stop and really think about that.

Do not be afraid. As Winston Churchill once said, "We have nothing to fear but fear itself." If you are a religious person, you may find it quite helpful to try saying a little prayer before you examine the remaining contents of this book. Ask ( ) to stay with you as you seek further truth.

Do you dare compare yourself with Jesus? You should and you must. Jesus never asked us to worship him. He asked us to be like him, to follow him. In other words, use him as our example. He was our perfect example for love and of the ability we all have to manipulate quantum energy. Jesus was able to say things like, "Let there be . . ." and immediately there was.

However, with the state of mind humanity is in at this time, I think that immediate gratification of our wishes would cause us to destroy one another. Nevertheless, Jesus did demonstrate that instant gratification of our wishes is possible, but until we get our minds pure, like Jesus, we had better move with caution as we learn to first love one another.

A most interesting book I had the pleasure of reading was *The Message of the Sphinx* by Graham Hancock and Robert Bauval. You should read about the extensive study they made. Like me, their findings will get you thinking.

The civilization that constructed the Giza complex more than some twenty-six thousand years ago, consisting of the sphinx and the great pyramids, were not of our time. What I mean is that they were from the last earth's procession.

Our earth proceeds through a procession brought on by a wobbling of the earth's axis, which takes around 25,920 years to complete its cycle. This wobble in the earth's axis is calculated to be as much as thirty degrees shift from dead center. That would cause enough variation from our sun to facilitate some major shifts in the earth's climate, hints, the onslaught and retreat of ice ages.

One should overlook the Egyptian pharaohs' inferior attempts to emulate the great pyramids' construction they admired so much, yet was never able to comprehend its purpose. The ancient civilization I am speaking of had the ability to manipulate quantum energy like nothing most of us have yet to even imagine today.

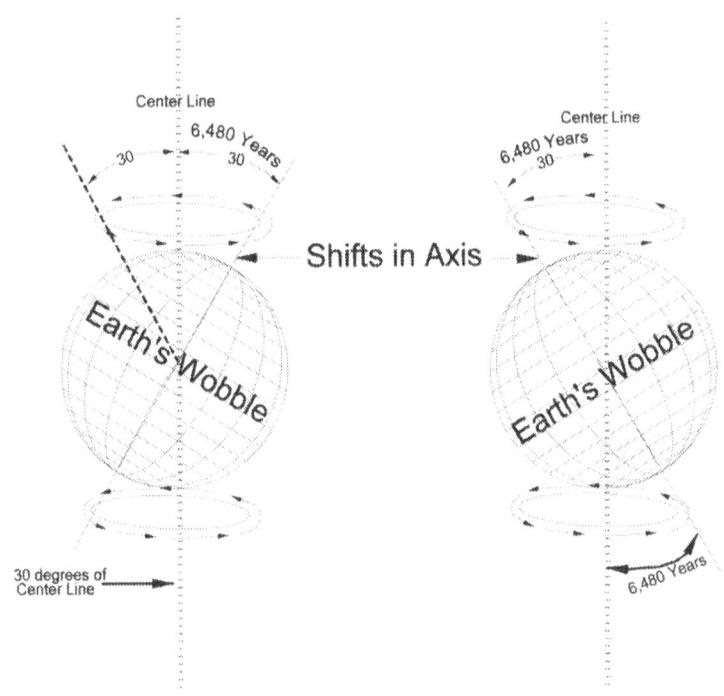

**The Earth's Procession**
**25,920. Years Cycle**

*The Earth's Procession*

What do the previous entry mean? It means that if we allow the keepers of the problem to stop us by shackling our minds, we will not be able to creatively and openly seek new ways in our thinking to solve the problems as we see them. Humanity will not move forward metaphysically with the necessary enlightenment needed to meet our future. We will be spiritually doomed as this procession comes to its end.

I say spiritually doomed because we are to develop technically as well as spiritually. At the rate we are moving, we will be traveling in deep space long before we ever realize our spiritual potential. At the end of the previous procession, the humans' spirituality outstripped their technology. At the end of their procession, everyone was so enlightened that they were able to transcend within a flash of light from the planet. Because such ideas were incomprehensible by those making the decision on what will go in the Bible or not, many things just had to be left out. You can, however, go to the universe for insight on this on your own.

Mostly everyone of the previous procession was enlightened. Noah, on the other hand, not having such enlightenment, received instruction for surviving and weathering the onslaught of water that came upon the earth. The task of reseeding the planet fell to him. Most of us know the story about the Great Flood, so let me continue by sharing my radical insight of the end time of the last procession.

This will require a completely open imagination on your behalf. A couple of centuries before the Great Flood when humanity was allegedly washed away, the universe started taking on a peculiar negative alignment; the magnetic poles shifted, allowing freed electrons of our planet to stream out into space at light speed. The people of the time recognized that by aiming these particles they could hitch a ride on these free electrons, affording them an out-of-body mode of transportation.

Those people were already practicing molecular manipulation by speeding up or slowing down the vibrations within atoms. The enlightened ones could rearrange atoms to fit their needs. With that ability came the awareness that they only had to transport their consciousness on to the electrons and aim for the stars. To get to other star systems, they needed a reliable, stable aiming device. The accomplishment of such an aiming device was through the construction of a great pyramid, perfectly aligned to the desired solar systems.

The electrons used for the transportation system were not just from any atoms; these electrons were stripped from around the nucleus of the atoms from their human bodies. Only the enlightened human beings knew the art of vibrating atoms into a desired atomic weight. With such knowledge, a rock could temporarily become as light as *aerogel* (a solid, consisting of 99 percent air).

Before long, their consciousness was able to move freely on an electron vehicle from the planet out into the universe, carrying with it all the knowledge it had accumulated as it was held together by electromagnetism in the human-mass form. This should hint to us that electrons can be knowledgeable.

Such electrons are capable of working together, forming whatever transportation device desired. They could travel as waves yet easily present themselves as particles when the transportation device chose. It could even change its charge and become what we know as proton and neutrons.

So how did this all come about? The people first learned to accept their quantum connection by allowing their spirit to manipulate the quantum energy. Soon every enlightened one's needs were met, and they were able to turn their attention to higher spiritual purposes. Before long, they no longer cared to stay in human form or on earth. This is what began the exodus from earth.

Would this mean that their spirit took control of their bodies as their consciousness gave way? Could the atoms of the human form still reconglomerate and work together after such a break up? Does spirit move into another direction after transformation is completed? Are these even good questions at this time? Perhaps these are questions for some future generation to find the answers to.

# CHAPTER 15

## The Latter Days

We are all on some mission as we journey through life. We are to be either the problem or the solution. The universal consciousness accommodates both in order to conclude its understanding of itself. Do not be resentful or discouraged with the situations you may find yourself in. See it only as your problem to be solved when you move through that station in your life. Find your mission and delight in being yourself.

Contrary to what I was taught, I have now come to the realization that ( ) does not discriminate; ( ) does not deal with what we deem right or wrong. If we are to discover ( )'s universes, then we, as ( )'s individual brain cells, must have the freedom to think creatively. We therefore are left with the task of coming up with what the rights and wrongs must be for ourselves as a society.

No matter what we decide as right and wrong, it becomes an inherited consequence in the reality for our children to endure. Even if our decision affords us some note of instant gratification, that gratification becomes our children's note to service.

In the past, the church was able to establish our sense of "right and wrong," while controlling the masses, by and largely, through fear. Through education, people are overcoming such fears. As we move to uplift all humanity, social consciousness must always be the forerunner. Any good or evil set forth in the universe becomes a part of the conscious of the universe. We must understand that polluting the universal consciousness with evil thoughts and fraudulent scams are forever captured, swindling the universal consciousness of harmony.

Every time we lie, cheat, and steal, we abuse the universal consciousness by corrupting its perception of itself. Whatever ( ) perceives in the perception we present will echo exponentially back into our society in a proliferation of more lying and cheating. Is this the way we want ( ) to perceive us? Is this the way we want ( ) to perceive itself?

The thoughts we think, dream, and scheme go out in waves of quantum energy. At the same time, these thoughts, while locating its like forms in our personal memories, can only congregate with like fellows in the universe as well.

We are all part of the universe, and thoughts can only attach themselves to like thoughts. We must all bear the resulting fruits. Be it good or bad, we are permanently marked as the recipients of any thought proliferation.

The universe is the storehouse for the creation of all possible realities. Recall that waves morph into particles, particles into matter, and matter into complex matter. Reality starts with a thought.

With so many different thoughts out in the universe seeking like thoughts in which to attach themselves, when a person decides to entertain a good or bad thought, they find that they can come up with many ideas they did not know they were even capable of formulating.

Before patting themselves on the back, they need to know that their party has been crashed by the infiltration of universal like thoughts; before long, these like thoughts have overrun their consciousness and are the governing force in their reality.

Have you noticed that people who steal think that everyone else does too? You can also see that honest people think that most people are honest. What is happening universally is that like thoughts are vibrating into like realities. The honest person is sensing the universe's like vibrations of other honest people, just as the dishonest can only flow in and with its like vibration. This can cause one to believe that everyone is like them or that everyone should see things as they do.

I have found that the Bible, along with other religious text, is full of incredible wisdom. Stay dilated, as I serve up some thought-provoking dessert for your imagination.

Back before Noah's Great Flood, humans of the past had amassed a great deal of wisdom. Some of this wisdom found its way into our civilization as religious text. This preflood civilization had little or no need for books or scripts, because they were able to comprehend each other's consciousness.

They had the ability to manipulate quantum energy into any solid state. With their minds, they could alter the electromagnetic state of any solid mass, simply by manipulating and altering its molecular structure.

They knew how to move their consciousness into any solid object and control its being.

If any of this seems strange, give it some time to germinate before passing judgment. Passing judgment is the best way to stop the universal flow of like thoughts into your consciousness.

Ask yourself, what would this mean if this were the case with that past civilization? Imagine what it would be like if we could tap into some of this past knowledge.

That knowledge is still out there in the universal consciousness. What's more, we are still wired up to handle this past knowledge. Why do you think we have such a large brain and only use 10 percent of it? That past civilization stretched our brain to this capacity. We are the offspring of that long-gone civilization through Noah. Yet, we are like babies with our big brain, and no teacher is capable of teaching us of its full uses.

Imagine the kind of development one of your offspring would receive if, in some peculiar circumstance, it was left as a baby with some primitive tribe. We need no stretch in imagination to quickly conclude that this baby, even with its inherent neurons to facilitate reading and writing, would not have much of a chance in developing above the comfort zone of the tribe's leadership.

We can apply the same thinking to us because, if not for the Giza complex sitting there in Cairo, we may not have ever questioned our true potential. Without a teacher from the past to teach us what we need to know to develop our full mental potential, all that we have learned to do is use enough of our potential to simply take care of ourselves.

It is not hard to see why some people, after achieving what they perceive as arriving, find themselves asking, "Is that all there is?" That's because once all of our basic needs have been satisfied, those once-dormant brain cells we inherited from this past civilization can then start to assert themselves.

As currents from the subconscious search throughout our memory banks seeking meanings for its existence, an occasional charge overspill will reach these uncharted higher brain cells. As these charges illuminate some of these past patterns of knowledge, the subconscious may not know what to make out of them. Remember, only brain cells of like hues can congregate.

So, if the said charge is so great as to cause brilliance in its illumination, the subconscious can be overwhelmed. Whatever patterns were in that locked-up part of those dormant brain cells will then become that subconscious' principal hue. The original hue that housed the subconscious' objective is then cast aside, as this new hue from the past takes command.

However, this new hue from the past has no objective to follow. It therefore randomly seeks out and views hues that are familiar to its hue. This would be the observation of quantum energy being manipulated in ways our conscious mind could find totally alien and frightening, simply because they have no like hue to match. Any spillage of such a subconscious hue into our conscious mind would not and could not congregate, because there is no previously exposed accommodating hue.

Nevertheless, here is one way people sometimes get a past-hue glimpse, distorted at best, into the achievements of this ancient archive. When the conscious mind is in slumber, the ancient hue sometimes roams unshackled and free in our subconscious, illuminating hues of its kind only.

As this ancient-hue activity is stored into our memory and observed on the slumbering-conscious level as a dream, our slumbering conscious, governed by the logics of today, does not know what to make of it. Remember, we can only interpret the hues of such dreams with the colors of our own familiar hues. Because of their hueless reception, such dreams may be described as strange and even as a bad dream.

There have been many innocent people victimized, even destroyed as demons, witches—the list is longer than I care to mention—because they were able to recall the ability in some ancient hue. What's even most distressing about those victimized people is that many of them adopted these negative labels and try to act them out in rebellion of the label itself.

Why should people be afraid of their own inherited ability? Why allow anyone in this day and time to shackle anyone's mind from remembering our past? If we are to get beyond our 10 percent capacity, we will have to stop listening to teaching of fear and learn to look within and remember for ourselves. Everyone should know that their brain is their inheritance, and no one has the right to scare them from the utilization of it.

It has always been the tactics of the cave dwelling, touch burners, to demonize and fall short of any light that is too bright for their eyes. Be, therefore, aware.

Not being taught in this higher state of consciousness, most of us have no matching hue that will link our consciousness to that ancient passageway. Some 90 percent of this once-developed brain is ignored and many times even suppressed with drugs as it is forced to continue its sleep. A few relics did manage to survive the Great Flood, and they have caused us to wonder. Nevertheless, we are still terribly misunderstanding these.

After the Great Flood, and millenniums later, people from Noah's lineage migrated into that once lush area and, over some few thousand years, started fashioning their society out of the veneration of the relics they found left in the area. They never quite understood its true message or meanings, even to this day.

Many people still believe that the great pyramids were tombs for the pharaohs. Not so, that Giza complex was never constructed for a pharaoh or a high priest. That complex was constructed to accommodate its people in their final enlightenment and transformation of their bodies and consciousness.

Long before the Egyptians' dynasties of their pharaohs were conceived, and long before they had formed a society that could even attempt having its people copy the pyramids, lived a great civilization millenniums ago, a society of a past procession.

Dilate your imagination now to the summer of this preflood society's procession, a time when their enlightenment really started to take hold. Ironically, this is about where we are now in our procession's wobble.

Their transformation did not take place until the late-winter harvest of their procession's end. They knew their procession's end-time thousands of years before it was to come, which gave them the time to understand and to facilitate the exact timing of their harvest. The sphinx was constructed as one of the precise monumental tools used in an aiming calculation.

To ensure proper alignment of their exodus from the planet, they cleverly constructed their sphinx in conjunction with the aiming device they would use to travel to other solar systems. Because the universe is forever changing, proper timing was critical. A mirror image of the sky, as it would appear to them at their harvest time, was also formulated in the construction of their launching complex.

This mirroring of the sky not only served as a time marker for them; it also told them precisely when harvest (launch) was to take place. The pyramids' construction and orientation provided the stable aiming device's need. Inside the great pyramid was where the final preparation took place for the launch.

To further emulate the sky, they even created the Nile River as a reflection of the many cluster of stars above. Well, when you have the skill, why not illustrate a portion of the sky as the Nile River as well?

Many of us, as we proceed into the summer of our procession, will start to see the way for our transition. To transform is to first gain a better perception of what we refer to as our consciousness. We must develop a perception that will allow our consciousness its ability to overwhelm standard beliefs, to break down and reconstruct the physics of our thoughts. This perception procedure will again gain its credibility through our understanding of the many high and low electromagnetic wave functions.

In other words, the consciousness crop the earlier people provided took on such an illumination that their bodies became unnecessary for their consciousness to survive. At harvest, the cells of their bodies broke down to its molecular structure; electrons were stripped away to serve as a vehicle for their consciousness to travel on, an "Out-of-Body System of Travel" (OBST).

Very little residue, by-product, was left due to the energy consumption in the fueling of the transformation process. The illumination from the process generated a flash so brilliant it had to take place in a boxed enclosure. The resulting electrons then moved into the launch chamber's opening as the transportation device.

How can we know when our harvesting will start to take place? By calculating through east, the mirroring of the Orion's Belt is directly over the great pyramids, in combination with the sphinx viewing Leo rising due east. Then, we will know that our harvesting time is upon us.

We could use the same solar alignment and follow our ancestors to the systems they chose, or we could build new pyramids to aim at completely different solar systems of our own likings and establish life as inspired by us.

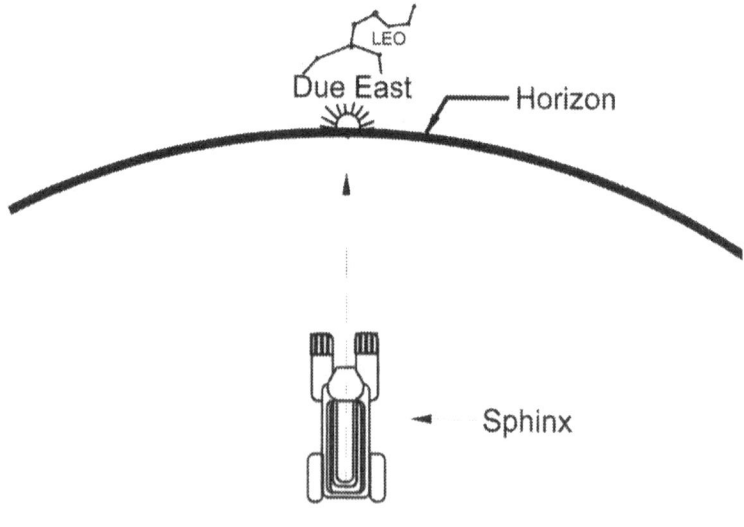

*Due East*

# Harvest Time

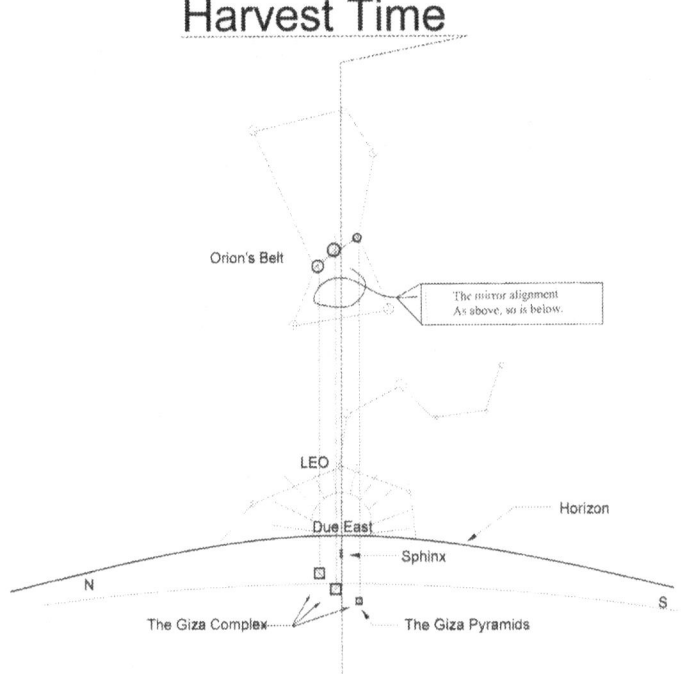

Orion's Belt

The mirror alignment
As above, so is below.

LEO

Horizon

Due East

Sphinx

N

S

The Giza Complex

The Giza Pyramids

*Harvest Time*

If we are still only working with less than 10 percent of our brain capacities, knowing the when and the how of the procession's ending will not matter at all. If by that time, the human race is not seeding the universe with colonies through our technology or has not developed the mental ability to make the harvest transition, well, let us not predict such an outcome.

As the essence of ( )'s consciousness aligns in its approach toward our planet (chapter 17 will discuss the PMS output), there will be such an overwhelming shock to our planet. Even the magnetic pole will shift around; the planetary electrons will become disarrayed. The power and brilliance will be so great that planetary upheavals will reign so extremely; no one will survive. Those who can will transcend into the brilliant will. Those who will seed the universe will have long been gone. Anyone not meeting either criterion will be doomed in the ruins of earth.

( ) desires both tracks to develop. Some of our offspring will go out in space in human flesh to develop more cells for ( )'s humanity. Others will become that pure consciousness that will join directly into ( )'s consciousness, while others will escape with the OBST transportation device. No matter what form our offspring takes, they will all be the procession's winter harvest. We should recognize that if this is the way ( ) has chosen to spread humanity and its consciousness throughout the universe, who are we to disagree? However it is done, its ( )'s prerogative.

This past knowledge is all still locked up in the UC, and with the proper like thoughts, we can access some of this knowledge of the past. The key word then is "like thoughts." But what were their thoughts like? It is very likely and understandable that we too, like the Egyptians, would misinterpret even the most basic concepts as we attempt in our struggle to find reason with only 10 percent of our perception's capacity in play.

If we employ our science and technology while seeking within ourselves, we will be able to better accept our findings. We will only find what we seek. It is there, within us all, a so-called divine inspiration, that spark of light.

We all somehow know when we are enlightened by this divine inspiration. We are all capable of being recipients in this connection of divine inspiration; no matter what philosophy someone has been taught, somehow, instinctively they know that they should not be killing another human being, don't they? Even when they have been acclimated to do so, something inside of them knows that it is not the way.

Noah and his family brought with them the ability of longevity from the past, along with some of the past wisdom for successful living. Most of that great mental knowledge was not brought along or was not taught to Noah. Was it that the mental capacity of that past civilization was taken into the whole, leaving a small handful of beings to weather the storm to start all over again, to develop technologically, slightly ahead of spirituality this time? Are we to seed the universe with technology at the forefront?

Allowing a civilization to develop such awesome technology as the splitting of the atom, while employing only 10 percent of their capacity for wisdom, would be the same as someone taking on the world with one hand tied behind their back. Well, if they were on loan from ( ), I suppose it could be done.

It is understood that the average teenager does not normally think with their higher brain, as the fully matured adult does. Therefore, we rarely extend any real responsibility to them. Humanity, on the other hand, while still spiritual teenagers, possesses the persuasive power in nuclear technology.

What's wrong with that picture? That picture should tell us that if we view nuclear technology as a persuasive power, we are all in big trouble. It means that we have a bully attitude, and we must quickly acknowledge that a bully only begets a bully.

We must begin doing all that we possibly can to elevate the consciousness of all humanity and not just ourselves. Realize that it takes only one unenlightened, misinformed individual to do the unthinkable. We can no longer ignore any part of our human family and hope for the best. We can accept our quantum connection or leave it to chance. Whatever we decide is going to be reciprocated throughout our universe.

When we think about past leaders who had the responsibility of spiritually leading their culture, we can understand that as those leaders came across information shared by cultures of the previous procession. They withheld that information from the public, out of their own lack of understanding and fear of their culture disintegration, or just a fear of change within itself.

They had great reservation about putting such awesome preflood power into the hands of their perceived subordinates. So today, we are left with only their interpretation of whatever the finding was. Hint: the many scrolls found but were left out of the Bible for example.

It was predicted, by Edgar Cayce, that a chamber under the right paw of the sphinx would be located and entered into during the end of the twentieth century. A team of Japanese scientists did go over to Cairo and did affirm the chamber through seismic wave.

It has become clear now that the keepers of the problem—for what they feel is for our own good—are not going to share the findings. Whatever is found will be kept hush, hush.

Do you think they are going to tell us about it? Not as long as they are afraid. Even though this information is for all of humankind, it is going to stay in the hands of a well-meaning few for as long as possible. It's going to be a blatant denial like that area 51 in Nevada once was.

But despair not, our ancient family, having seen well into the future, predicted that would be the case. So, what they did was store the archive in a place where all humanity can have access to them. They created the sphinx to convey a story, a story once understood and cannot be denied. The human head with a beastly body must also be seen with its positioning into the earth. We know that when the sphinx is left unattended over a period of time, the earth reclaims it.

When Napoleon invaded Egypt, he came upon this great head of a human sticking out of the sand. Once unveiled, the body of a beast was revealed.

Here is the simple story revealed through this time capsule known as the sphinx. The human head tells us that we are nothing more than a powerful beast without the knowledge stored in our inherited head; it is only what is in our head that makes us different than the beast. It further tells us that we must look high into our head to be enlightened.

We must look to our head for the enlightenment of our bodies, for the enlightenment of the human body. For years, we have been looking in, under, and around the sphinx for the archives of humanity, when it has been staring us right in our face all the time. Somehow, Mr. Edgar Cayce got the idea that the archives were stored in the paws of the sphinx. What he missed was that if you stood on the right paw of the sphinx and looked up, you would be looking to the right side of the sphinx head, the creative side.

Before anyone goes out and smash open the right side of the sphinx's head, let's understand that it is in the right side of our head that the archives are stored, in the head of each and every one of us. Like the sphinx, the earth will reclaim our animallike body if we cannot acknowledge our inherent archives.

What can we interpret in the sphinx facing the east? Well, for one thing, it is a time marker to mark the east, as it will appear at the end of each procession. It is also an extraordinarily clever way of telling us about accessing our inherent knowledge. The only way to know is to seek for yourself; as you seek, you will learn some proper hues for the ancient memories to congregate into consciousness.

The sphinx is a monument of the earth as the motherhood for humanity in conjunction with the other three pyramids; it conveys the time of harvest season, which occurs at the end of each procession. Some may have heard of it as "the end times."

As I surfed the net, I came across the sphinx photo. Out of curiosity, I tried taking the nose of one of my mother's photo and affixed it on the sphinx's face. I did the front as well as a side view. Suddenly, this "sphinx"-structure face not only makes sense; it looks like the face of many, if not most Negro women. Voila! The sphinx looks like my mama.

*Mother as the Sphinx*

No, it's not the sphinx's face; it's my mama's face!

It's understandable that the pharaohs would have had trouble with that broad nose from the past and, therefore, removed it and replaced it with their keener nose to accommodate the philosophy of its developing culture. There are still old chisel markings, which support the deliberate removal of her nose. They felt this had to be done to facilitate a replacement design, which would look more like their pharaoh's nose.

However, this monument was fashioned from one continuous limestone rock. In trying to affix the pharaoh's attached nose with their technology at that time, the replacement could not stay on for long. There was no way it could have withstood the test of time.

So what does this all prove? It proves that we cannot always rely on what anyone else tries to convince us of about the past and about what they want people to believe in as the truth. If we are to become seekers of truth, we must learn to go into the archive for ourselves. Only you can go into your archive and learn the truth about what is important to you.

Imagine this: The so-called sphinx is not only in celebration of earth's motherhood; the people of the past procession knew that we would one day have to reestablish the fact of our feminine origin. In spite of our stubbornness, that sister is still speaking to us.

They knew that because of the effects caused by the earth's procession; by the time our consciousness reached the summer of our procession, the truth about our ancient past would have been so distorted out of fear, jealousy, and prejudice that we would not be able to access our archives.

For one to access their archives for themselves, they must approach the right hemisphere in a relaxed state of mind and have with them a properly stated question. A proper thought is in the form of a "like thought" or idea that is already hued up in your ram memory of consciousness. If you have properly done this, you can start your enlightenment journey.

It is also very interesting to see how the ancients have made it next to impossible for anyone to access their archive, if they are unable to make those first steps of discarding negative attributes from their mind. It is not a difficult task; we only need to first know and accept "Who's your mama?" In order for anyone to share in a family's inheritance, they must first show that they are of the family.

When we can accept the answer to "Who's our mama?" we can then accept who our sisters and brothers are. When we accept "Who are our siblings?" we can then accept "Who's our family?" When we can accept our family, we accept a shift in consciousness. This consciousness shift is what facilitates the hues that are capable of congregating in our family's archives—that is, when we are standing on the right paw of the sphinx and looking up into her right hemisphere.

Let us make an example. If you were looking for a way to concentrate light, like lasers, you could not ask the proper questions without some scientific knowledge of light first. My point is this: one could not comprehend light as a source of, let us say for example, propulsion, if one did not understand the physics involved in light. Therefore, the thoughts of such propulsion could not enter one's mind. But even if they just heard of it from others, they still could not recognize the physics in the solution.

So, here's my point, even though all of the answers we could ever seek are already in our archives, we cannot access them until we know how to ask with a like-thought hues. Paradox? Not really, this is more of a safety net. In the mental state we are now in, what would we do with such awesome power and knowledge?

We still have not been able to decide and understand our purpose in life. We have not even realized that we need a shared mission statement for humanity. Without a mission statement, how does everyone know humanity's aim? It is not our Creator's intention for us to wander aimlessly through life until we destroy ourselves. If the unenlightened mind was able to access the archives, what would they do with such overwhelming, awesome power? The answer is clear to most of us: they would use it to control others.

We saw what happened when the unenlightened got their hands on the divine inspiration of splitting the atom. Would it surprise you to learn that the unenlightened minds have created enough warheads to destroy humanity several times over and over again?

Now why would any sane person want to create anything that would destroy humanity even once? That's what happens when we allow fear to outflank our good humanity senses; we allow the keepers of the problem to simply keep the problem; we allow the unenlightened to shackle humanity into their cavemen mentality. Now you can see why you are not to discuss your archives with anyone. Some great thinker once said, "He who knows does not talk; he who talks does not know."

Accessing the archives is for the enlightened ones. It is therefore counterproductive to try and explain this book to the unenlightened. Either they are able to understand it for themselves or they are not ready to get it at this time in their life. We can see what happened when some enlightened minds explained the splitting of the atom to the unenlightened. If someone does not understand their quantum connection for themselves, let them reread the book.

Here is another reason why you cannot explain to the unenlightened. Imagine everyone on the planet living in total darkness, with the exception of a beam of light shinning from their forehead as their only source of light. This light is their only reality, and it shines only for them. Some may have twenty-five-watt bulbs, which only allow them to see just what's in front of them, while others may have one-hundred—or even one-thousand-watt bulbs, illuminating far out in front and to the side.

On the other hand, the person leading you, your teachers or perhaps even your employer, may be operating behind only a fifty-watt bulb, yet they are insisting that you stay behind them. One can see why the fifty-watt may become upset when the 150-watt tries to point out something in the shadows beyond their vision.

# CHAPTER 16

## Our Family Reunion

I think it is counterproductive to hate anyone because of his or her differences. Skin color, for example, is only due to regional differences. To say you hate someone because they are brown, yellow, white, or black is to say that you hate your ancestral parents. It was through our ancient parents that brought us to where we are today. We now know that humans migrated from the bushes of Africa, and I am referring to the bushes of Africa before the previous procession as well.

The passageway of any migration trail continuously would be broken by the onslaught of ice ages, severing all contact for generations upon generations. As the earth moved through its 26,900-year procession, our skin had to change colors in order to facilitate the amount of sunlight in the given region. The isolation of any small group, in any different latitudinal areas for a few hundred generations, would cause each individual in that group to share in the same appearance as well as culture.

These earlier migrating humans, not having the ability to record their past, would certainly experience the raising of eyebrows, not to mention the fear, whenever an ice age receded and they ran into their unfamiliar cousins. However, with our understanding now, we can all rethink the negative emotions of hate some of us harbor for each other. Other than our ignorance of ourselves, this hate has absolutely no basis.

Let's take a look into the future for a moment. As we enter outer space to do our universal calling, populating other planets, we will evolve and even mutate to facilitate our lives on other rocks we will someday find ourselves being born on. We will engineer our genes to facilitate the needs of our planetary exploration. These human beings will undoubtedly look much different from us today. Fortunately, we are now able to keep records of our past. The future humans must be able to celebrate our differences in brotherly love.

We have only one race, and that is our human race. We must all learn that "we are one." Moving beyond this impasse will be a real sign that we have grown up. We have a lot more to worry about than our skin color or our religious differences, and if we keep wasting our procession's time, none of us will have a skin to worry about or a religion to believe in.

I would like to see the whole earth set one weekend aside every ten years as the "family-reunion time" for the whole planet. This tradition should be established and encouraged throughout space as we expand our species. In this way, we shall never again lose contact or forget that we are all from the same "Mama."

Imagine all of humanity preparing for this up and coming weekend of celebration. This would be a business stimulus for several million people, not to mention the strong sense of sisterhood or brotherhood this celebration would create. Again, there is that win-win situation. This seems like a workable way in which humanity can stay bonded.

Without such a tradition, we would be leaving future star wars to chance. As we better understand the nature of the universal electromagnetic forces, we will be able to hurdle whole planets at one another. I am sure that if we still hate one another in the future, we will have long destroyed ourselves before gaining such fantastic abilities.

We already have the means to demolish our fragile planet; it's just that the level of negative energy has not been able to gain critical mass to take over the universal consciousness. This is a state where the whole planet loses all sense of reasoning. One can see this demonstrated when a group of people gets swept up in the mentality of a mob consciousness.

So there you have it; we can continue letting others do our thinking for us as they, unknowingly, lead us down the path of hate and distrust, or we can learn to do our own thinking and start celebrating humanity as our own family.

# CHAPTER 17

## Finding One's Mission in Life

To live one's whole lifetime on earth without freedom, just doing what others would have you do for them, must be a disappointment to ( ). If one does not have the freedom of their own thoughts, what would they have? How can ( ) benefit through your life if you do not have control of it? If ( ) is to experience the universe through you, how can you contribute to that experience if you are not able to think for yourself? You must be free to think for yourself. Thinking for yourself is the only way you can think for ( ). Coordinate your garments on your easterly voyages.

This universal consciousness is the consciousness of ( )—that is, consciousness as demonstrated through humanity's servicing of their spirit.

Like snowflakes, we are all created with different talents and abilities. It is vital that one learns to enhance what he or she has within them. One should never lose time trying to be "like Mike," when they have been given the ability to be a writer or a dancer, or whatever. It is also counterproductive for one to spend their precious lifetime envying the talent of others.

Envy is a trait that usually starts its development at childhood between siblings, thus making it one of the tougher traits to break. Usually this trait takes its roots when a child notices one of the siblings displaying abilities they do not share. Unless the parent takes proper intervention, the child will take this trait into adulthood and his community. Trying to "keep up with the Jones" is not the quantum-connection way; one must be oneself.

But what does it mean to be oneself? It means to be the person that spirit has brought them into consciousness, not the person our parents and others thought best for us. Yet, that may be the mechanism of us getting on track with our spirit.

I see nothing wrong with a parent putting their child on a goal, even if the child changes it when they get older. I have required my children to come up with a goal by the time they reach the seventh grade, or I put them on one. I put them on one because they need to have a reason to do all the work required of them through school. This has worked like a charm.

Some people are multitalented and many times have problems trying to land on one thing because they are able to do so many things well. However, being able to do many things well does have its advantage by giving insight into many professions as they go through life.

In finding one's mission in life, one could also ask themselves these simple questions: Does what I want to do feel right within my spirit? Does it excite my spirit? Does it make me happy? Can it provide a livelihood? Does it enhance humanity in some way? Is it within the law? If the answers are yes, then start doing that thing at once, part time if necessary, even if you cannot make money at it right away.

If you have found the thing you should be doing in life, you should be willing to do it on the side with pleasure. Eventually, that thing you love will provide you a handsome living if you persist.

My mission in life, I used to think, was to ask and answer my SAQ. Sir Isaac Newton must have been another person who asked and answered his SAQ. Imagine how the people of his time must have looked at him when he wanted to discuss why an apple would fall to the ground, when everyone knew that it just does. Most people were happy with "it just does."

When a spirit brings a person to life as a truth seeker, then seek they must, even when they themselves may not understand that they are such. They may only know that they have some SAQs that demand some answers.

It is through spirit that our consciousness was brought to life, and it is through our consciousness that we create the reality that ( ) observes and experiences an understanding of us within itself. Therefore, we are free to create whatever we like. So, be cognizant that whatever we create becomes ( )'s perception of its universal self.

With all due respect to the renowned physicist Stephen Hawking, I wish to apologize for having the audacity to differ with some of his theories on the universe, for it is through Dr. Hawking's as well as Dr. Albert Einstein's theories that I have been afforded the insight into my universal model. Although I now know that my mission in life is to track down what we refer to as consciousness and most of all "spirit," I found it necessary, if I am to accomplish my mission, to give thought to, and model the universe for, myself.

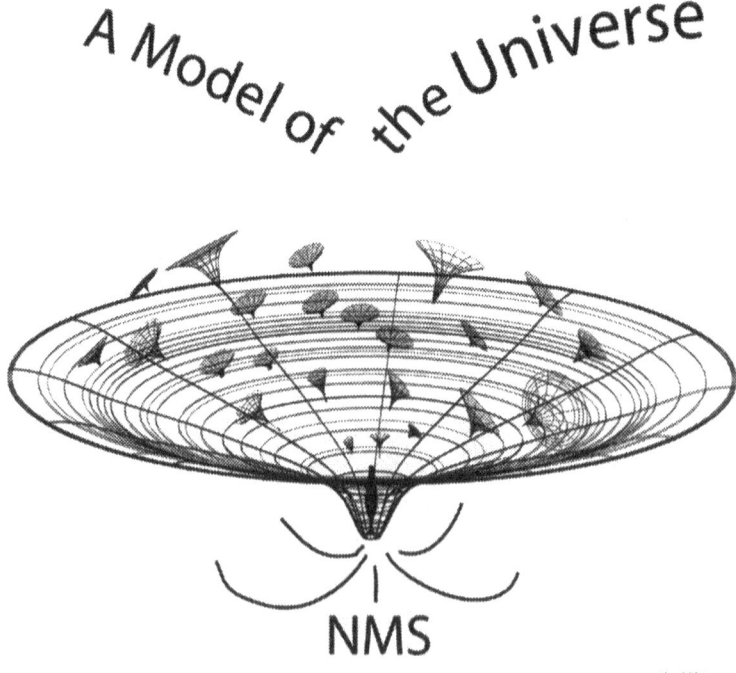

A Model of the Universe

NMS

Moo2004

In my model, the universe has an imaginary shape of an elliptical funnel. The ellipse portion is visible through our eyes, but the funnel's drainage area must be imagined as an electromagnetic force, pulling the ellipses into its center drain or black hole. The funnel or the helix shape is the natural formation all bodies of matter must assume, as their direction of motion is influenced and opposed by the electromagnetic inward pull. We can observe such helix phenomenon when they reveal themselves throughout the earth, as a water flow and in air turbulence. This is the same force that created all of the solar systems and galaxies.

I further hypothesize that the universe will not come to a cold, dark end, and never shall all matter be sucked back into a vast, cataclysmic gravity well. The reason is simple; the universe is continuously regurgitating itself through the properties that sprout out from the black hole.

Galaxies are teaming with crushing black holes, which serve the function of recycling the body of the universe. As masses fall prey to the black hole's crunching system, all the life gets squeezed out of it. "What life?" one may ask. The force that held the mass in its form is the life.

This then brings the force to the forefront for clarity—a body in motion attracted by the electromagnetic field of a larger body, somewhat as we early learned in "like thoughts." However, if the motion of the smaller body can maintain a high-enough speed, the resulting relationship can only be one of an orbit. This then becomes a standoff between motion and gravity.

What then would be the result if the electromagnetism could and was squeezed out of the orbiting relationship? Does such a thing happen, and what benefit is there, if any? The answer is yes, it does happen, and the benefits are astronomical.

It is the function of the black hole to free this spirit by crushing it and spurring it out into the universe, an anti-by-product (dark matter?) which I call the NMS (neutral magnetic source). NMS must make room for itself by pushing its way out into an already crowded universe, causing everything else to expand away, due to its antisocial characteristic. Nevertheless, we can all appreciate why one would be a bit grumpy after being embraced, fondled, and molested by such an event horizon.

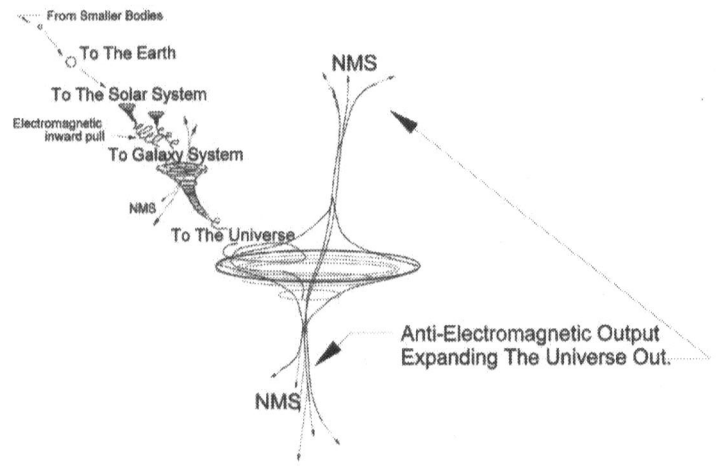

The Illusion we call Gravity

*NMS, Expanding Back into the Universe*

With this NMS transformation going on throughout the universe, to accommodate NMS, more room is required. Thus, we detect an expanding universe.

Accepting the universe as expanding through this NMS must then suggest that the universe would be overfilling with this NMS and, thus, forcing the visible universe to be pushed out.

With this NMS functioning against the gravitational function, the two opposing waves cause new particles to be formed out of the helix effect. These loose particles then become the new raw materials for the formation of new bodies into solar systems and new galaxies.

Gravity must therefore be "nothing" more than a result of the helix-functioning effect: as smaller bodies are captured by larger ones, the moon falls to earth; the earth falls to the sun, the sun to galaxy, galaxies to the universe, and universes to the *infiniverse.*

My theory obviously discards any wormhole theory as just wishful thinking. That does not mean that we cannot find a way that will allow us to leap into other galaxies. What it does mean however is that we are not properly hued up at this time to conceive the way. This is just another reason why all human beings should have the freedom to think creatively. An impoverished mind is not free.

My mission in life is to track down "spirit and consciousness"; your mission in life is within you as well. That is the only place where one should look for their mission. Your spirit holds the answer. It is only by way of your spirit that the answers will come. When one has managed to align their consciousness in harmony with their spirit, they will know that they have found their mission in life by the many dopamine showers that mission will bring.

With the completion of this book, one should then be able to live as a cognitive quantum-connected person. If you find that you would like to discuss such things as: How do thoughts travel? How does prayer work? How does desire become reality? and What is consciousness? These questions may still seem like just some SAQ to some, but there are real physics involved in each. If you have the freedom of unshackled thoughts, I invite you to pick any of the previously mentioned SAQ and join our Quantum Connection Society forum, a place where you can find other like minds, a place were you can share your quantum-connection success stories. I too will join in as often as I can at *www.thequantumconnection.com.*

# APPENDIX 1

## A Goal for Humanity

Every successful corporation has clearly stated goals and objectives. Such goals allow everyone in the corporation to know where the corporation wants to go. These goals also give the public some insight for buying into that corporation or not. As we are now entering the midpoint of our earth's procession, it is time we state some goals for ourselves (humanity). The human mind works best when it knows and understands the reasoning of what is ask of it.

Here are my suggestions: before anyone attempts to set any goals for humanity, we should first debate on what kind of humanity we want to develop. Are we going to move out into outer space and seed the universe with human cells? This would be the catalyst that justifies the investment to create a highly technical, space-driven world economy.

We also need to work out ways for every human being to be free to grow spiritually. The universe has, in spite of what appears to be turmoil, an order to which it is evolving. The same laws that govern gravity, fission, and fusion also dictate the radical notion that our consciousness is free to soar as well. Somehow, we all have an innate desire to fly like the birds. We have this desire because the universal laws require that of us. We, therefore, will one day soar from the restriction of our bodies.

One thing is certain: the procession's end will find us unprepared if we cannot organize ourselves and work together as one race. There is also a good chance that the procession end will not find us here at all, if we do not evolve above the conflicting religious doctrines we now share. ( ) has given us free will; we must allow each human being to think for themselves as well. No one can follow their own spirit if they are not given the opportunity to think above the dogma they have been indoctrinated with.

*The Quantum Connection Theory* advocates thinking for oneself, seeking the truth through one's spirit, not through the spirit of someone else. Most of all, know that you need only to hold fast to the ( ) idea.

# APPENDIX 2

## A Prayer of Acknowledgement

Before we make another prayer, we should first recognize what a prayer is and why it works. Prayers are nothing more than an attempt to facilitate or affect a favorable outcome in one's reality. Many of us go about this by first personifying and then praising ( ) into processing a wave function of their desire. Personifying ( ) gives one a way of focusing their quantum energy. The reason people go about it that way is because they have been taught to think so little of their inherent creative abilities. They are unable to conceive that ( ) emanates through all of us and that through that power we can all achieve what we believe.

How prayer works will be further addressed in my next book as I dig deeper into consciousness. What we need to know now is that we are ( )'s ambassadors, and we have the power to create in ( )'s behalf. We are to be creative with our freedom as we contribute to ( )'s corporation.

## The Prayer

Ubiquitous ( ), now equipped with the understanding of my quantum connection, I will no longer act like an out-of-control selfish brain cell. I will not be a part of any out-of-control, brain-eating cancer cells. I believe that I can have all that I desire through full acknowledgment that I am a part of the brain cells that service you, ( ). It is only through your desires that I am your eyes, ears, limbs, and consciousness, created by you to fulfill your desires and wishes. It is only through my exploitation of this arrangement that my desire is granted.

I will do everything I can to cease doing harm to any other part of your humanity. I will look to you, ( ), for inspiration as to how I should conduct myself. Most of all, let me not have shame in seeking help, should it ever be revealed to me that I have some social deviance that brings harm to another human being.

# BIBLIOGRAPHY

A.R.E., Inc. Official Cite of Edgar Cayce's ARE-Association for Research and Enlightenment. ® A.R.E., Inc. 2002 Edgar Cayce Readings © By the Edgar Cayce Foundation <http://www.edgarcayce.org/>

Hawking, Stephen. *The Universe In a Nutshell.* New York: Batnam, 2001.

*Holy Bible: Red Letter Edition.* Iowa Falls: World Bible Publishers.